The Art and Science of Teaching Agriculture: Four Keys to Dynamic Learning

THE ART AND SCIENCE OF TEACHING AGRICULTURE: FOUR KEYS TO DYNAMIC LEARNING

M. SUSIE WHITTINGTON

RICK RUDD

JACK ELLIOT

PDF AND EPUB FREE ONLINE AT: https://doi.org/10.21061/teachagriculture

Virginia Tech Department of Agriculture, Leadership, and Community Education in association with Virginia Tech Publishing and the Open Education Initiative of the University Libraries at Virginia Tech

Blacksburg

Suggested citation: Whittington, M. Susie, Rick Rudd, and Jack Elliot, ed. (2023). *The Art and Science of Teaching Agriculture: Four Keys to Dynamic Learning.* Blacksburg: Virginia Tech Department of Agricultural, Leadership, and Community Education. https://doi.org/10.21061/teachagriculture. Licensed with CC BY-NC 4.0 https://creativecommons.org/licenses/by-nc/4.0.

Publisher: This work is published by the Virginia Tech Department of Agriculture, Leadership, and Community Education in association with Virginia Tech Publishing and the Open Education Initiative of the University Libraries at Virginia Tech

Virginia Tech Department of Agriculture,
Leadership, and Community Education
214 Litton-Reaves Hall
Blacksburg, VA 24061 USA

Virginia Tech Publishing,
University Libraries
560 Drillfield Drive
Blacksburg, VA 24061 USA

Open Education Initiative,
University Libraries
560 Drillfield Drive
Blacksburg, VA 24061 USA

Peer review: This book has undergone editorial peer review.

Accessibility statement: Virginia Tech is committed to making its publications accessible in accordance with the Americans with Disabilities Act of 1990. The Open Education Initiative is committed to continuous improvement regarding accessibility. The text, images, headings, and links in the PDF and HTML versions of this text are tagged structurally and include alternative text, which allows for machine readability. Please contact openeducation@vt.edu if you are a person with a disability and have suggestions to make this book more accessible.

Publication cataloging information:
M. Susie Whittington, Rick Rudd, and Jack Elliott, editors
The Art and Science of Teaching Agriculture / M. Susie Whittington, Rick Rudd, and Jack Elliot
Pages cm
ISBN 978-1-957213-71-2 (PDF)
ISBN 978-1-957213-66-8 (Print)
ISBN 978-1-957213-70-5 (ePub)
ISBN 978-1-957213-72-9 (Pressbooks | https://pressbooks.lib.vt.edu/teachagriculture)
URI: http://hdl.handle.net/10919/116050
DOI: https://doi.org/10.21061/teachagriculture

1. Agriculture—Study and teaching—Textbooks.
   Art and Science of Teaching Agriculture: Four Keys to Dynamic Learning, The
   Title S531.A78

Cover design & illustrations: Kindred Grey

To our parents who were our first teachers;

To our children who taught us to be masterful teachers by teaching us the art of patience, negotiation, tenacity, and unconditional love;

To our formal and nonformal teachers from cradle to today who were forever teaching us to love teaching before we were even aware of their influence across our life spans; and

To our talented authors who through their collection of thoughts, best practices, strategies, and techniques graciously share with readers a glimpse of the teaching environments into which they breathe life year after year.

—the Editors

CONTENTS

FOREWORD

Keep Growing: How Inspiring Leaders Grow Themselves, Their Educators, and Their Learners

A great leader is able to inspire a teacher to grow and develop from being mediocre to good, good to great. This process requires a leader to be purposeful and intentional in their pedagogy; that's what *The Art and Science of Teaching Agriculture: Four Keys to Dynamic Learning* is all about.

As identical twin sisters and colleagues, Marla and I have been fortunate to share experiences in the same profession of education. That's what we hope for you, too.

Although our areas of leadership have differed somewhat, what we have found is that our outcomes have been similar due to our connected, culturally competent leadership style. We believe that cultural competence is foundational to the art and science of teaching.

Becoming culturally competent is a journey. It is not a place where you *land and stay*. As situations arise you move back and forth on a continuum. You don't become good at it and that's it. You strive to always grow to be better.

Being a culturally competent leader means first, as leaders, that we come to the educational environment with a firm sense of who we are, and that unless we are reflective about our own identity and how it creates a lens through which we view the world, we will not be able to honor the identities of the learners and educators we serve.

In order to develop great educators, a leader must listen deeply to those around them. We have found that being purposefully aware and responsive, and listening to the broader learning community in ways that ensures all community members feel valued, requires consistent work, but the results are certainly worth it.

In addition, inspiring leaders must (1) have organized goals for their teachers that are visionary, goal-driven, innovative, and motivational, (2) possess professional skills such as being knowledgeable and communicative and acknowledging the success of others, and (3) possess personal attributes such as enthusiasm and integrity and being an active listener, respectful, and approachable. These qualities are no different from what a leader wants to see in their educators, therefore these traits must be modeled.

A great, inspiring leader must keep their educators motivated so they inspire their learners. This is a teacher's most important role and an essential part of their learners' overall success.

But keeping educators motivated can be challenging. Some suggestions for educational leaders are to:

1. **Respect them.** Use positive, respectful language at all times and really listen to their concerns, thoughts, and ideas.
2. **Nurture greatness.** By giving your teachers new ways to become better at what they do, you are constantly reminding them why they entered the profession in the first place. Hold high expectations for them and they will perform. Consider investing in more effective professional development to help you in this quandary.
3. **Give them a voice.** The easiest way for teachers to lose motivation is when they have been quieted or not allowed to speak at all. Invite teachers to the table whether it be for representing their department or subject area at board meetings, or for giving their opinion about proposed strategies and visions. Even if they are not directly involved in particular meetings, having an *open-door* policy just to listen creates opportunities for educators to feel like they matter, even if they are not directly involved.
4. **Encourage collaboration.** When teachers are given the opportunity to work together, this can significantly impact their motivation. More seasoned, experienced teachers may be recognized for their most successful teaching strategies. Younger or less experienced teachers can be given validation that what they are doing in their learning environments is working.
5. **Be available.** Help your educators as much as possible by being available when they have a difficult situation or heavier workload.
6. **Say "thank you."** Recognize the *little things* that teachers are doing. Everyone wants to feel valued. When we take the time to say *"thank you,"* we are helping teachers stay motivated. You might also want to set aside time during staff meetings for other teachers to recognize their peers for recently doing something positive.
7. **Let them know you understand the stressful times.** Try not to overload educators during those busier times of year. Try not to overburden them with new initiatives or professional development during report-writing periods, exam periods, preparation for parent-teacher conferences, or preparation for community-based events, to list a few examples. If there is any way you can lend a hand during these periods, whether it be helping with the county fair exhibit, cafeteria or recess duty, or room setup, then try to do so.
8. **Be resourceful.** A leader who wants to help their staff become great must be available to offer suggestions along with resources. While answers aren't always at your fingertips, you must know how to help teachers move from the mediocre stage of teaching to good and even masterful! This requires leaders to create and continually develop a plethora of ideas and actual useful tools.

This textbook is one of those tools. *The Art and Science of Teaching Agriculture: Four Keys to Dynamic Learning* is written by professors who believe in both the art and the science of teaching. It is written from hearts who are passionate about the holistic approach to learning environments. It is written by educators who advocate for students to share partnership in the teaching and learning process. And it is written by great leaders who combine decades of experience in educational environments in which they have intentionally and strategically sought to inspire, motivate, and grow toward mastery in their art and science.

We hope you, like us, find yourselves fortunate to thrive in the profession of education.

Enjoy *The Art and Science of Teaching Agriculture: Four Keys to Dynamic Learning.*

Monica Marsh and Marla Marsh

August 2023

INTRODUCTION

The Art and Science of Teaching Agriculture: Four Keys to Dynamic Learning is a collection of thoughts, best practices, strategies, and techniques assembled from among the best teaching professors in agricultural communication, education, and leadership in America. The editors have assembled the authors' mastery work as a way for all of us to assist our aspiring teachers in grasping the fundamentals of solid, basic, time-tested formal and nonformal teaching. *Four Keys to Dynamic Learning* simply means that we believe there are four keys to teaching that must be present in the learning environment for learners to achieve at their highest levels: Laying the Foundation, Connecting with Students, Designing Instruction, and Applying Learning. Embedded within these dimensions are the unique aspects of both the art and the science of teaching that are necessary for engaging learners in the teaching and learning process.

The Art and Science of Teaching Agriculture: Four Keys to Dynamic Learning is brought to life in this book by the selfless contributions of these authors, who, as researchers, have studied the teaching and learning process, have implemented what they have learned, have modified their planning, delivery, and assessment, and who are continuing their quest for mastery in their formal and nonformal learning environments.

Historically, across the country, preservice teacher education in the agricultural sciences has shared common foundational principles. Traditionally, these shared principles have been, in part, perpetuated through a common textbook used in many preservice teacher education methods courses in higher education. Therefore, an objective of *The Art and Science of Teaching Agriculture: Four Keys to Dynamic Learning* is to provide a multiperspective resource for educators to use in preparing those who seek to teach others. The goal is to positively influence the caliber of methods of instruction globally.

The text is organized around *Four Keys* that need to be considered when preparing educators to plan, deliver, and assess learning: Laying the Foundation, Connecting with Students, Designing Instruction, and Applying Learning.

Role of the Editors

We, the editors, combined our collective more than 100 years in education to read drafts, offer insights, and challenge authors to think beyond traditional classrooms and into both formal and nonformal learning environments. Therefore, it was important to us that the book's use in education not be limited to traditional preservice classroom teacher preparation. One approach to accomplishing this was choosing verbiage such as learning environments, educators, and learners to capture our spirit and intent in speaking to both formal and nonformal education. We also read authors' examples through a lens of reaching our formal and nonformal audiences so that all educators using the book would find examples for use in workshops, seminars, field experiences, laboratories and other environments in which learners learn.

Throughout the process we stayed connected to each other as the chapters evolved. We read the document holistically to assure *one voice* for the book, while maintaining the unique, individual voices that constitute the heart of *The Art and Science of Teaching Agriculture: Four Keys to Dynamic Learning*. We also read through the drafts looking for three elements woven through the fabric of the chapters: diversity and inclusion, assessment, and effective teaching.

Finally, to aid in our quest for a broad use of the book across formal and nonformal education, students in a graduate teacher education program created sample personas of people they believed would be the audience for the book. The personas served as a resource to assist the authors in focusing their writing and messaging. Examples included but were not limited to the following personas.

Personas

A first-year faculty member with a 100 percent teaching appointment and no previous teaching experience beyond serving as a teaching assistant in graduate school	
Basic demographics	Hispanic male who is thirty-two years old.
Where are they teaching?	A small liberal arts college in West Virginia.
Who are they teaching?	Undergraduate students of impoverished backgrounds.
Their general teaching practice	A combination of lectures and laboratory experiences. He incorporates different pedagogical techniques other than standard practices that include more activities, higher student engagement, but lacks access to technology beyond PowerPoint.
What questions would they like to have answered?	• How does he get on the same page as his students from diverse backgrounds? • How could he tailor his teaching approach from large to small classes? • How does he write assessments? • How does he design a complete course from scratch?

A first-year international master's student	
Basic demographics	A female who speaks English as a second language, mostly fluent in English, but worried about connecting with students. She is excited to acquire teaching experience.
What are they teaching?	A basic, required course in their department at a land-grant university. Teaching a large class of more than 100 students. Some students are engaged, while some just have to be there. It is possibly the first time the students are exposed to this material. In the past, the class has been taught using lectures. The current teacher is only thought of as average by the students.
Their general teaching practice	They want to be more engaged but are concerned about power balance in class. The current teacher has been teaching the same way for a long time. They are teaching using mostly lectures.
What questions would they like to have answered?	• How does she ensure that her language is not a barrier or an issue with the students? • Are there stylistic or cultural differences between her experiences and those of her students in the United States? • Is there anything she should know about various strategies? • How does she have students respect her as a teaching figure and not as a peer?

Mrs. Bean, who has no prior teaching experience	
Basic demographics	Female from an underrepresented population, is vertically challenged in stature, and a bit of an introvert.
What are they teaching?	A minor class with prerequisites.
Their general teaching practice	She tries to look confident and competent. She wants students to be engaged. She wants to have an interactive classroom.
What questions would they like to have answered?	• How does she act to be confident when teaching? • How can she motivate students with low self-motivation? • How should she incorporate different activities in the classroom?

Community College Instructor with a 100 percent teaching appointment	
Basic demographics	A transgender female, white, adult (twenty-eight to forty years old).
Who are they teaching?	Nontraditional students from diverse cultures and ethnic backgrounds, some international students, some students from a local high school. Some students have low socioeconomic status and some are working full-time jobs, including night shifts. There are students in the class with families.
What are they teaching?	Some of her courses are offered at night and on weekends. The day classes are offered for those students who work night shifts.
Their general teaching practice	She is energetic, uses PowerPoint, and is very structured. She also uses online course management software.

New faculty member who has recently completed their doctoral program	
Basic demographics	Female, early thirties, with minimum previous teaching experience.
What is the context in which they are teaching?	She both teaches and conducts research. She teaches undergraduate and graduate students. She has no budget for teaching activities and is creating a new course. There is resistance from faculty colleagues to implement new teaching techniques.
What questions would they like to have answered?	• How does she negotiate with peers and old faculty? • What should she do to get funding for teaching? • How does she implement class activities? • How does she find and use new and affordable and accessible teaching technologies? • How does she build an environment of trust in the classroom?

Who Are the Authors?

We were purposeful and deliberate in our selection of the authors. We invited authors who were known across the country as being excellent in their ability to plan, deliver, and assess teaching and learning. We invited authors who were, at the time of invitation, nontenured, new assistant, or associate professors in universities across the country. We intentionally combined coauthor teams across disciplines, backgrounds, and areas of expertise. Finally, we sought authors who had achieved an earned recognition for the content we were asking them to espouse in their chapter.

Unique Framing of the Chapters

As context for users of this book, our guidelines to authors, in an overly simplified approach, was for them to capture on paper "the best class session you have ever taught on the topic you are capturing in your chapter." Our *ask*, therefore, included:

- A practical scenario to which the intended audiences could relate, thereby aiding in setting the context for learning the content of the chapter.
- A clear objective for the outcome of learning the content, thereby providing clarity of expectation for the learner.
- A science- and theory-based grounding, thereby providing evidence of rigorous scrutiny of the content prior to adoption, but not with the intensity in writing of a journal publication. Instead, nonformal, relatable writing was sought.
- An application of the content, thereby offering learners opportunities for active engagement with the material.
- An assessment of the achievement of the objective, thereby giving learners and educators a reflective process for deciding the level of content attainment.

In addition, the editors asked the authors to use theory as a framework and to provide anecdotes from personal experiences. The authors were asked to close each chapter with reflective questions for the learner. These reflections are meant to be broader than the scope of the objectives to offer more holistic reflection over the content of the chapter. We also asked the authors to include a glossary as well as a list of commonly used resources for the audience to use in their teaching.

Finally, we asked the authors to weave every chapter through the tapestry of the following elements:

- Diversity and Inclusion
- Evaluation
- Effective Teaching

The Chapters

Part 1: Laying the Foundation

1. The Discipline of Agricultural Education

"The Discipline of Agricultural Education" provides a foundation for learning throughout the textbook by exploring what agricultural education is, why it exists, who it serves, and how it is delivered.

2. Psychology of Learning

"Psychology of Learning" digs into how learning occurs by introducing some essential learning theories and how teachers can apply these learning theories to help students who have unique learning styles and needs against the backdrop of societal norms in the United States. Theories tackle students' identities. Diversity and inclusion are discussed, alongside theories of learning.

3. Principles of Teaching and Learning

"Principles of Teaching and Learning" examines evidence-based concepts related to how students bring in and process information and explores foundational truths about teaching that can help students grasp new information, think critically, and retain content.

Part 2: Connecting with Students

4. Learning as Problem Solving

"Learning as Problem Solving" lays out problem-based instructional strategies that can help create unique opportunities to connect our content back to issues and problems in our local communities as well as statewide, nationwide, and worldwide.

5. Inclusive Teaching

"Inclusive Teaching" is intended to help educators begin the process of purposefully planning courses, Future Farmers of America (FFA) opportunities, and Supervised Agricultural Experiences to be inclusive of all student cultures and identities that may be present in the local community.

6. Dynamics of Teaching

"Dynamics of Teaching" aligns the art and science of implementing change in behavior, environment, and personal factors to enhance teaching.

Part 3: Designing Instruction

7. Planning for Effective Instruction

"Planning for Effective Instruction" introduces the process of curriculum design including writing learning objectives, mapping curriculum, and designing lesson plans for impactful teaching.

8. Delivering Content with Technology

"Delivering Content with Technology" explores how to mediate content delivery, classroom management, and student assessment strategies through appropriate technology platforms.

9. Assessing Agricultural Education

"Assessing Agricultural Education" investigates how agricultural education can be equally valuable as high stakes testing scores based on rigorous evidence and data from the variety of student driven learning and projects that are synonymous with the content we teach.

Part 4: Applying Learning

10. Applied Leadership Development through FFA

"Applied Leadership Development through FFA" highlights the experiential learning opportunities inherent in America's largest agricultural youth organization. We explore the structure of this intracurricular component of the complete agricultural education model, as well as the many contextualized learning experiences instructors can use in challenging diverse learners to combine technical knowledge and skill with interpersonal know-how to provide inclusive growth opportunities for all learners.

11. Supervised Agricultural Experiences

"Supervised Agricultural Experiences" explores the history, benefits, areas of concern, and development and implementation factors that impact student engagement in this individualized educational learning experience.

12. Effective Use of the Agricultural Laboratory Environments to Support Student Learning

"Effective Use of the Agricultural Laboratory Environment to Support Student Learning" introduces types of laboratory environments and provides guidance for planning for laboratory instruction.

Summary

The Art and Science of Teaching Agriculture: Four Keys to Dynamic Learning is more than a book; it is a collaboration of highly skilled and respected teaching professors sharing their thoughts, best practices, strategies, and techniques about a topic they love. The book offers educators a solid foundation for building confidence in planning, delivering, and assessing the depths of the variables inherent in a learning environment. The editors wish for readers the ability to teach dynamically by designing and managing learning environments using what we believe to be four keys to good teaching: Laying the Foundation, Connecting with Students, Designing Instruction, and Applying Learning.

Keep teaching well!

Susie, Rick, and Jack

ABOUT THE EDITORS

M. Susie Whittington

M. Susie Whittington, PhD, is a Distinguished Professor of Food, Agricultural, and Environmental Sciences and Executive Director of the Second-Year Transformational Experience Program at the Ohio State University. She was the first woman inducted as a Fellow in the American Association for Agricultural Education.

Dr. Whittington taught Methods of Teaching for twenty-five years. She created and taught for ten years a university general education course, Toward Cultural Proficiency. Online she teaches Advanced Methods of Teaching. For more than twenty-five years, she directed a research project, *Improving the Cognitive Capacity of Students by Fully Engaging Professors in the Teaching and Learning Process*, which received national recognition from her professional organizations. This body of work took Susie and her husband to India, Taiwan, and Kenya to build capacity in student-centered teaching. Dr. Whittington is the junior author of *Methods of Teaching Agriculture* and *Toward Cultural Proficiency*. Her teaching has been awarded the USDA National Teaching Excellence Award, and the North American Colleges and Teachers of Agriculture's Teaching Award of Excellence. She received the Josephine Sitterle Failer Award for Outstanding Service to Students by the Ohio State University Alumni Association. Dr. Whittington earned her BS, MS, and PhD from the Ohio State University.

Rick D. Rudd

Rick D. Rudd, PhD, is the Community Viability Chair of Excellence and Professor of Agricultural and Extension Education (ALCE) at Virginia Tech. Rudd served as ALCE Department Head from August 2006–July 2019. He served as Interim Associate Dean and Director of Virginia Cooperative Extension and Professor in 2009–2010. He received his PhD from Virginia Tech in 1994. He earned his bachelor's degree and master's degree from the Ohio State University.

Rudd is currently serving as the Director for Workforce Education and Development in the Center for International Research, Education and Development at Virginia Tech. In this role, Rudd is responsible for seeking and securing international projects that utilize Virginia Tech's expertise in agriculture, workforce education, youth development to cultivate social and economic advancement. Rick recently accepted the role of Faculty Principal in the Leadership and Social Change Residential College at Virginia Tech. This role is an addition to his other responsibilities. He and his wife live in the residence hall with more than three hundred students!

Dr. Rudd is working on community viability models that can lead to regenerative communities, community centered youth development, as well as methods for planning, designing, delivering, and evaluating international agricultural and workforce education and development programs. He has directed and codirected over ten million dollars in funded scholarship efforts. Rudd has international experience in Russia, Senegal, Tanzania, the Dominican Republic, South Sudan, Jordan, Palestine, and Honduras.

Jack Elliot

Jack Elliot, PhD, is the Regional Director for Africa for the Borlaug Institute for International Agriculture and Development. He serves on the USAID Higher Education Learning Network Steering Committee and leads the Council of Research and Evidence (CORE). He is a professor in the Texas A&M Department of Agricultural Leadership, Education, and Communications (ALEC) where he served two terms as the Department Head, providing leadership for almost 1,500 students, and 80 faculty and staff. As president of the International Association for Agricultural and Extension Education (AIAEE), he organized the first non-US annual conference in Trinidad and Tobago. He established and became the first AIAEE journal editor. Elliot received his BS and MS in Agricultural Education and Agricultural Economics from Washington State University. He earned his PhD in Agricultural Education from the Ohio State University. He was awarded the FFA National VIP Award in 2023.

Currently, he leads four USDA/FAS International Agricultural Education Fellowship Programs. Prior to his career in academia, Jack was a dryland grain farmer, cattle rancher, and operated a custom harvesting business in Belt, Montana, for seventeen years.

ABOUT THE AUTHORS

Chapter 1: The Discipline of Agricultural Education

Wendy Warner

Wendy Warner, PhD, is an Associate Professor of Agricultural Education in the Agricultural and Human Sciences Department at NC State University. Warner's journey from being a high school agricultural education student in rural Ohio to teaching middle school and high school agriculture in urban Orlando, Florida, shaped Warner's interests in urban agricultural education and in engaging diverse audiences through classroom and laboratory instruction, leadership development, and experiential learning. Warner earned a BS from the Ohio State University and PhD from the University of Florida, both in agricultural education.

Aaron J. McKim

Aaron J. McKim, PhD, is an Associate Professor of Agriculture, Food, and Natural Resources (AFNR) Education at Michigan State University. McKim's research focuses on enhancing interdisciplinary learning throughout educational experiences. McKim's passion for combining disciplines within learning experiences began as a middle school and high school educator in Indiana. Today, McKim works with preservice and in-service educators in Michigan and beyond to illuminate learning across disciplines via the context of agriculture, food, and natural resource systems. McKim earned a BS in agricultural education from Purdue University, and an MS in agricultural education and a PhD in science education from Oregon State University.

Chapter 2: Psychology of Learning

Hui-Hui Wang

Hui-Hui Wang, PhD, is an Associate Professor in the Department of Agricultural Sciences Education and Communication at Purdue University. Wang also has a joint appointment with the Department of Curriculum and Instruction. Wang's research interests are integrated science, technology, engineering, and mathematics (STEM) instruction and inquiry-based and project-based teaching through agriculture, food, and natural resources (AFNR) in both formal and nonformal settings. Wang's research focuses on investigating and building sustainable and interdisciplinary STEM through an AFNR teaching and learning model for K-12 educators and students. Wang also works with the United States Agency for International Development Feed the Future Innovation Lab for Food Safety projects to assist project leadership teams to address gender and women's equality. Wang earned a BS in biology and wildlife conservation from the National Pingtung University of Science and Technology, MA in environmental education from Southern Oregon University, and PhD in curriculum and instruction in science education from the University of Minnesota.

Summer Odom

Summer Odom, PhD, is an Associate Professor in the Department of Agricultural Leadership, Education, and Communications at Texas A&M University. Odom teaches undergraduate and graduate courses related to leadership including personal leadership education, leadership program facilitation, public leadership development, and learning organizations. Odom primarily conducts research related to college students and effective pedagogies for teaching leadership including design of high-impact practices that are conducive for learning leadership. Odom has been involved in design-based research using design thinking as an innovative practice for teaching student skills in the areas of problem solving and teamwork. Odom earned a BS in food science and technology, MS in agricultural education, and PhD in educational human resource development from Texas A&M University.

Chapter 3: Principles of Teaching and Learning

Kasee L. Smith

Kasee L. Smith, PhD, is an Associate Professor at the University of Idaho in the Department of Agricultural Education, Leadership, and Communications. Smith instructs courses related to teaching and learning, social science research, and verbal communications. Prior to working at the postsecondary level, Smith taught high school agriculture and advised an FFA chapter in Spanish Fork, Utah, for eleven years. Smith earned BS and MS degrees in agricultural education from Utah State University, and a PhD from Texas A&M University.

Chapter 4: Learning as Problem Solving

Carla Jagger

Carla Jagger, PhD, is an Assistant Professor of agricultural education in the University of Florida/IFAS Department of Agricultural Education and Communication where Jagger specializes in agricultural education and teacher preparation. Jagger teaches coursework related to teaching methodologies in formal and nonformal environments, specifically in laboratory settings emphasizing problem-based instruction. Many of Jagger's research interests focus on building learning environments for all learners, especially underrepresented student populations. Jagger earned BS, MS, and PhD degrees in agricultural education from the Ohio State University.

Chapter 5: Inclusive Teaching

Stacy K. Vincent

Stacy Vincent, PhD, is the Director of Undergraduate Studies for Agricultural Education at the University of Kentucky. During Vincent's thirteen-year career as a college professor, his research has consistently focused on finding a voice, providing a voice, and giving a voice to underrepresented and marginalized youth in agricultural education. Vincent was previously a high school teacher for seven years and has over fifteen years of teaching experience at the postsecondary level. Vincent's research and expertise seek to assist teachers and students in gaining the competence levels of knowledge, awareness, behavior, and skill toward educating multiculturally different learners. Vincent earned BS and MS degrees in agricultural sciences from Murray State University, an MA in educational leadership from University of Louisville, and a PhD in agricultural education from the University of Mississippi-Columbia.

Donna Westfall-Rudd

Donna Westfall-Rudd, PhD, is the Director and Senior Scholar for the College of Agriculture and Life Science Graduate Teaching Scholars program and the Undergraduate Director for the Department of Agricultural, Leadership, and Community Education at Virginia Tech. Westfall-Rudd works with undergraduate students planning careers as secondary agricultural education teachers and doctoral students planning to pursue faculty positions with teaching appointments. Embedded in all Westfall-Rudd's work in teaching and learning is an emphasis on developing teachers as advocates for all underrepresented people in agriculture. Westfall-Rudd's educational program planning work also includes collaborations with women agricultural community educators and university faculty in Senegal. Westfall-Rudd earned BS and MS degrees in agricultural education and a PhD in education all from Cornell University.

Chapter 6: Dynamics of Teaching

Tobin Redwine

Tobin Redwine, PhD, is the lead learning analyst at Vivayic, Inc., and an instructional assistant professor at Texas A&M University. Redwine is an internationally awarded instructor and researcher with emphasis on agricultural communications and youth development in international settings. Redwine brings "a love of music, dad jokes, and cross-cultural application" to learning sessions. Redwine has trained youth secondary agriculture teachers in Uganda, Ghana, South Africa, and all over the United States. Redwine earned an M.S. in agricultural communications from Texas Tech University and PhD in agricultural leadership, education, and communications from Texas A&M University.

Chapter 7: Planning for Effective Instruction

Amber H. Rice

Amber H. Rice, PhD, is an associate professor in teacher education at the University of Arizona in the Department of Agricultural Education, Technology and Innovation. Rice is "passionate about teaching and learning and believes that curriculum development is the cornerstone to quality, effective teaching practice." Rice has vast experience in creating curriculum for learners in various formal and nonformal contexts and across experience levels and believes that knowledge of the learners is the single most important consideration for creating meaningful curriculum. Rice earned a BA in agricultural education and MS in career and technical education from the University of Kentucky and a PhD in agricultural education and leadership from the University of Missouri.

Matt Mars

Matthew M. Mars, PhD, is an associate professor in leadership and innovation at the University of Arizona in the Department of Agricultural Education, Technology and Innovation. Mars is passionate about helping students develop the skills and knowledge necessary for advancing new ideas through innovative strategies and practices. Innovation is inherently abstract and requires students to confront a range of complex principles, concepts, and models. Curriculum development and planning is critical to setting a strong foundation from which students can learn, grow, and evolve into effective change agents. Mars earned a BA in sociology/anthropology from Utica College of Syracuse University, MEd in counseling/human relations from Northern Arizona University, and a PhD in higher education from the University of Arizona.

Chapter 8: Delivery Content with Technology

OP McCubbins

OP McCubbins, PhD, is an Associate Professor of Agricultural Education at Mississippi State University. McCubbins's classes focus on teacher preparation and directing learning experiences. McCubbins has taught agriculture at the secondary and postsecondary levels and loves and uses technology frequently to transform the learning environment for students. McCubbins's research focuses on the use and impact of immersive technology for teaching and learning in and about agriculture. McCubbins earned a PhD in agricultural education from Iowa State University.

Annie Specht

Annie Specht, PhD, teaches agricultural communication at the Ohio State University. Specht's classes focus on visual and verbal communication, and Specht loves introducing learners to their future profession through technology-assisted experiences such as virtual tours and guest speakers. Specht can usually be found watching movies and TV shows (for science!) when not training future communicators. Specht earned BS and MS degrees in agricultural communication from the Ohio State University, and a PhD in agricultural communications and journalism from Texas A&M University.

Chapter 9: Assessing Agricultural Education

Tiffany Drape

Tiffany Drape, PhD, is an agriscience educator, lifelong learner, and an associate professor at Virginia Tech. Drape's research interests include equity and access in agriculture and the life sciences. As an immigrant, a farmer's daughter, and an educator, these attributes inform much of the work Drape does as an educator and in assessing agriscience and CTE programming. Drape earned a BS in communications and an MS in agricultural education from Cornell University, and a PhD in agriculture and life sciences from Virginia Tech.

Chapter 10: Applied Leadership Development through FFA

Laura L. Greenhaw

Laura Greenhaw, PhD, is a former high school agriculture student, FFA member, and secondary ag teacher. Currently, as Assistant Professor of Agricultural Leadership at the University of Florida, Greenhaw is passionate about developing individual leaders and the leadership capacity of the agriculture and natural resources industry holistically. Greenhaw's favorite part of being an educator is helping students become critical thinkers and lifelong learners. Greenhaw earned a BS in agriculture, an MS in agricultural education, and PhD in agricultural communications and education from Texas Tech University.

Paula Faulkner

Paula Faulkner, PhD, is a former middle school educator. Currently a Professor of Agricultural Education in the Department of Agribusiness, Applied Economics and Agriscience Education at North Carolina Agricultural & Technical State University, Faulkner instructs undergraduate and graduate students in agricultural education. Faulkner possesses a great passion for preparing students, especially those identified as underrepresented, for postsecondary education and careers in the Food, Agriculture, Natural Resources, and Human (FANH) sciences and STEM areas. Faulkner earned a BS in agricultural technology and an MS agricultural education from North Carolina Agricultural & Technical State University. Faulkner earned a PhD in agricultural and extension education from Pennsylvania State University.

Chapter 11: Supervised Agricultural Experiences

Eric Rubenstein

Eric Rubenstein, PhD, is an Associate Professor of Agricultural Education in the Agricultural Leadership, Education, and Communication Department at the University of Georgia. Rubenstein's research focuses on the development and implementation of Supervised Agricultural Experiences programs and utilization of experiential learning in agricultural education. Rubenstein started his career in agricultural education as an agricultural education student and FFA member and subsequently taught middle and high school agriculture in Pennsylvania. Today, Rubenstein works with preservice and in-service teachers to engage all students in the agriculture education programs in Georgia middle and high schools. Rubenstein earned a BS in agricultural education and extension education from Pennsylvania State University, and an MS and PhD in agricultural education and communication from the University of Florida.

Chapter 12: Effective Use of the Agriculture Laboratory Environment to Support Student Learning

Hannah H. Scherer

Hannah H. Scherer, PhD, is an associate professor and extension specialist in the Department of Agricultural, Leadership, and Community Education at Virginia Tech. In this role, Scherer supports teaching and learning of agriculture in formal and nonformal settings through research and application of contemporary models for STEM education. Scherer has always had a passion for science and making sure that everyone has the opportunity to learn science in a way that aligns with their interests and passions and has done this as a science museum educator, high school science teacher, and university professor supporting agricultural education. Sherer earned a BA in geology from Macalester College, and a PhD in geological and environmental sciences from Stanford University.

Annie Specht

Photo courtesy of the Department of Agricultural Communication, Education, and Leadership at Ohio State University

Matt Mars

University of Arizona College of Agriculture and Life Sciences

M. Susie Whittington

Provided by M. Susie Whittington, The Ohio State University

Donna Westfall-Rudd

Virginia Tech College of Agriculture and Life Sciences Communications

Kasee L. Smith

Printed by authorized permission of the author

ACKNOWLEDGMENTS

The editors are indebted to Anita Walz and her team at the Open Education Initiative at Virginia Tech for their time, talent, and expertise.

Editorial Team

Anita Walz, Project Manager and Managing Editor

Anita Walz is Associate Professor, Assistant Director of Open Education, and Scholarly Communication Librarian in the University Libraries at Virginia Tech. She received her M.S. in library and information science from the University of Illinois at Urbana-Champaign and has worked in university, government, school, and international libraries for over 20 years. She is the founder of the Open Education Initiative at Virginia Tech and the managing editor of over twenty open textbooks adapted or created at Virginia Tech, many of which may be found by visiting https://vtechworks.lib.vt.edu/handle/10919/70959. She has provided overall planning, project encouragement and coordination, day-to-day management, coaching, problem-solving, oversight, and post-production marketing for this book.

Kindred Grey, Illustrator and Production Manager

Kindred Grey is the OER, Graphic Design, and Digital Publishing Specialist in the University Libraries at Virginia Tech. She joined University Libraries after receiving her B.S. in Statistics and Psychology from Virginia Tech in 2020. Kindred has contributed to over twenty open textbooks, providing technical support on layout and design. Her main focus is publishing open textbooks that are visually appealing, accessible, student oriented, and technologically advanced. She illustrated this book and also led production efforts including accessibility and formatting for the PDF, EPUB, print, and HTML versions of the text.

INSTRUCTOR RESOURCES

Please use the instructor interest form if you are an instructor who is reviewing, using, or adapting *Epidemiology* for a course, or if you have created any materials related to this book and would like to share them publicly.

Navigate to the book's main landing page to access:

- Links to multiple electronic versions of the textbook (PDF, EPUB, HTML)
- A link to order a print copy
- A link to the errata document
- A link to report errors

If you are interested in making your own version of this book, check out the book's license and information on best practices for attribution.

About Open Educational Resources (OER)

OER are free teaching and learning materials that are licensed to allow for revision and reuse. They can be fully self-contained textbooks, videos, quizzes, learning modules, and more. OER are distinct from free online resources in that they permit others to use, copy, distribute, modify, or reuse the content. The legal permission to modify and customize OER to meet the specific learning objectives of a particular course make them a useful pedagogical tool.

About Pressbooks

Pressbooks is an open source, web-based authoring tool based on WordPress. Virginia Tech personnel use it to create and adapt openly licensed course materials. Pressbooks offers an intuitive editing environment for educators interested in customizing this book.

About This Book

Helpful URLs

Main landing page (DOI): https://doi.org/10.21061/teachagriculture

CC BY-NC 4.0 license: https://creativecommons.org/licenses/by-nc/4.0/deed.en

Instructor interest form: https://bit.ly/teachagriculture_interest

View errata: https://bit.ly/teachagriculture_errata

Report an error: https://bit.ly/teachagriculture_errors

Pressbooks (HTML): https://pressbooks.lib.vt.edu/teachagriculture

Best practices for attribution: https://wiki.creativecommons.org/wiki/Best_practices_for_attribution

1. THE DISCIPLINE OF AGRICULTURAL EDUCATION

Wendy Warner and Aaron J. McKim

Setting the Stage

As a primer for this chapter, let's discuss *why* agricultural educators are essential. The following excerpt, shared by a former student during their agriculture teacher's retirement celebration, helps illuminate the importance of agricultural educators.

> *"Before taking Ag classes, high school was uninspiring. I remember my first class, 'Fundamentals of Agriscience,' it was so different from every other class at school. The more you let me experience, however, the more I felt at home in the ag room. Sure, the ideas and concepts you taught were interesting and important, but the way you treated me is what I remember the most. Somehow, you created a culture where I could be authentically myself at school. In your classes, I found my passion for animals, plants, and the environment. With every new FFA experience you pushed me toward, I became more prepared to lead. As my SAE expanded, so did my confidence that a career in agriculture was right for me. As I look back on everything you have done for me, every experience you provided, I know you changed my life for the better. You led me to my passion for agriculture and gave me tools to be successful in life. I am forever grateful for everything you have done for me and for all the students you impacted throughout your career."*

Before reading on, take a moment to reflect on the legacy we hope to achieve as agricultural educators. As we reflect on these goals, write down the impact we want to make throughout our careers.

Being agricultural educators gives us a platform to positively transform the lives of our students, the communities in which we work, and various agricultural systems. We hope this book helps us discover the impact we can have and empowers us with the skills and knowledge to achieve our desired legacy as agricultural educators.

Objectives

The purpose of this chapter is to introduce our readers to agricultural education. More specifically, the objectives are to:

- Discuss the role of instruction, Future Farmers of America (FFA), and Supervised Agricultural Experience (SAE) in agricultural education.
- Distinguish between different types of agricultural education.
- Explore various perspectives as to why agricultural education exists.
- Consider who is served by agricultural education.
- Explain how agricultural education is delivered.
- Provide an overview of the textbook.

Introduction

This chapter lays the foundation for the textbook by providing an overview of agricultural education and prompts consideration of what agricultural education is, why it exists, who it serves, and how it is delivered. Throughout the remainder of the textbook, we will learn about and reflect on the four keys to a dynamic learning environment. In part 1 of the textbook, we will be encouraged to think about how both the psychology and the principles of teaching and learning influence how agricultural education is delivered. In part 2, "Connecting with Students," we will consider how the use of problem solving, inclusive strategies, and the classroom environment allows for a connection with our learners. As we progress through part 3, the focus will be on "Designing Instruction" through intentional planning, delivery, and evaluation. We highlight "Applying Learning" in part 4 when describing the inclusion of leadership development opportunities through the FFA, the promotion of experiential and work-based learning through SAE, and the reinforcement of research and inquiry-based learning in the various laboratories found in agricultural education.

Overview of the Discipline of Agricultural Education

What is Agricultural Education?

Instruction

Instruction in each school-based agricultural education (SBAE) program has uniquely adapted to meet the needs of learners, the community, and local agricultural systems. These adaptations yield a rich variety of SBAE programs throughout the country, in each state, and within different regions of each state. Two examples of this diversity can be found in Michigan and North Carolina. SBAE instruction in the northern part of Michigan, including the Upper Peninsula and the northern part of the Lower Peninsula, focuses on natural resource management and forestry. For programs located in the southern half of the state, SBAE instruction focuses more broadly on animal science, plant science, agribusiness, and leadership. In North Carolina, new curricula have been developed and implemented to meet the interests and needs of a changing student demographic as have the opportunities available across different parts of the state, especially with the growth of agricultural programs in suburban and urban areas. Recently, courses in sustainable agricultural production were introduced with a focus on the increasingly complex world of producing enough food and fiber to meet the growing world demand and at the same time maintaining ecological balance and conserving our natural resources (North Carolina Career and Technical Education, 2020).

Diversity in SBAE instruction, however, is not reserved for geographic differences. In fact, when we look at the historical evolution of SBAE, we see differences in SBAE instruction due to changes in the learners, communities, and agricultural systems. A 2017 publication by McKim et al. organized the history of agricultural education in the United States into three categories, starting with the "Early Years of Agricultural Education." This first phase included agricultural education occurring in the United States before the Smith-Hughes Act, which formalized school-based agricultural education. Agricultural education in this time period was primarily done through father-son apprenticeships with some private schools teaching agriculture. One school emerging during this time was the Tuskegee Normal and Industrial School, opened in 1881 by Booker T. Washington, which provided former slaves or descendants of slaves with a combination of academic, social, and industrial training (Croom, 2007). The Hatch Act, which catalyzed experiment stations, shifted instruction in SBAE to focus on the distribution of scientific agricultural knowledge being gained at experiment stations to rural communities throughout the country. Instruction during the "Early Years of Agricultural Education" was a unique blend of scientific agriculture and vocational (i.e., career) preparation.

The second phase of agricultural education in the United States was named the "Smith-Hughes Years of Vocational Agriculture." This phase started in 1917 with the passage of the Smith-Hughes Act. During this time period, agricultural education instruction was characterized by SBAE teachers preparing learners for careers in farming. As education continued to evolve, tracking learners into either academic or vocational routes became the norm, further distinguishing SBAE as a space for vocational preparation. Further, societal changes encouraged SBAE instruction to prepare learners for a broader range of careers beyond farming, including mechanics, horticulture, and natural resource management. SBAE instruction during the "Smith-Hughes Years of Vocational Agriculture" was, as the name would suggest, vocational in nature.

The third phase of agricultural education has been named the "Current Focus on Agricultural Science." This phase started in 1988 with the publication of *A Nation at Risk: The Imperative for Educational Reform* and *Understanding Agriculture: New Directions for Education*. Noting additional core academic credits being required for graduation from public high schools, the *Understanding Agriculture* publication suggested a shift from vocational preparation to agricultural science to provide learners taking SBAE coursework with science credit. Since 1988, SBAE instruction has trended toward the illumination of science concepts within agriculture, food, and natural resources. Further, growing emphasis on science, technology, engineering, and mathematics (i.e., STEM education) has expanded the interdisciplinary nature of SBAE. The "Current Focus on Agricultural Science" phase, therefore, is defined by an emphasis on core subject (e.g., science) illumination while maintaining a foothold in vocational preparation through Career and Technical Education.

The diversity of SBAE instruction found geographically and historically illustrates a discipline defined by adaptation and evolution. As we look to the future, the nature of instruction within SBAE is uncertain; however, we can be sure that instruction will continue to evolve to meet the needs of learners, communities, and agricultural systems. As future agricultural educators, *we* will be in the driver's seat as the discipline navigates change and continues to move forward toward a better tomorrow.

FFA

What do you think our lives might have been like if we lived in the early 1900s? In 1911, more than 80 percent of children lived in the country, so chances are we too would have lived in a rural area. How might that be similar or different from where we currently live? How might our schooling have been different? The importance of agriculture could have influenced some of our coursework with a primary focus on learning specific skills for farming such as planting and harvesting crops or raising livestock.

Consider some of the extracurricular clubs we participated in during high school. Were we members of the student council or robotics club? Perhaps we were involved in the math club or drama club? However, if we lived at the turn of the century, we may have found ourselves participating in a corn club or a tomato club at our schools. Just like the student clubs we were a part of, these clubs brought youth together with a common interest to participate in activities and competitions.

You might be asking yourself, "What initiated the start of corn clubs?" A lack of adult interest in farmers' institutes in Macoupin County, Illinois, spurred the president of the county institute, W. B. Otwell, to distribute corn to 500 boys who grew it and competed for prizes at the following institute in 1900. This type of exhibition grew so quickly in popularity, that the following year, 1,500 farmer boys submitted entries and a statewide exhibit in 1904 drew participation from 8,000 boys (True, 1929). Agricultural clubs, such as corn clubs, also grew in popularity amongst rural schools as a way to encourage the study of agriculture (Davis, 1912). In 1902, A. B. Graham, who was school superintendent of rural schools in Springfield, Ohio, started what was known as a "boys' and girls' experiment club" (Reck, 1951, p. 12). Graham introduced learners to scientific concepts and agricultural skills such as soil testing, ropework, and using a microscope. Over time, Graham continued to develop and implement programs related to corn production, vegetable gardens, and flower gardens (Reck, 1951). By 1909, there were more than 10,000 boys enrolled in the competitive corn contests, and in 1910, over 46,000 boys were connected with corn clubs (True, 1929).

In 1909, Marie Cromer, a young rural schoolteacher in South Carolina, learned of the influence of corn clubs on crop yields. She proclaimed the need for a similar opportunity for farm girls and by the next year had initiated a girls' tomato club (Engelhardt, 2009). As part of the club's activities, girls were expected to plant one-tenth of an acre of tomatoes and provide all the effort during the growth and the canning of the crop (Evans et al., 1916; Pound & Moore, 2012).

The passage of the Smith-Hughes Act in 1917 led to the development of student organizations for rural youth that had an emphasis on agricultural production but also provided opportunities for personal growth and development (Croom, 2008). Walter S. Newman, who served as the State Supervisor of Agricultural Education in Virginia, grew concerned that farm boys were becoming disinterested in farming and eventually deciding to leave the farm. In an effort to give the boys a "shot in the arm," he suggested the idea of starting an organization for boys studying agriculture (Yeatts, 1954). During a meeting in September 1925 with Harry Sanders, Edmund Magill, and Henry Groseclose, Newman expressed his concern about the feelings of inferiority often experienced by farm boys. In response, he suggested forming an organization that would allow rural youth the opportunity to "develop confidence in their own ability and pride in the fact that they are farm boys" (Yeatts, 1954, p. 18). Around a wooden table at Virginia Polytechnic Institute, the Future Farmers of Virginia was born. A couple of years later, in May 1927, the New Farmers of Virginia was started at Virginia State College under the guidance of Dr. H. O. Sargent and G. W. Owens. This organization, which was similar to the Future Farmers of Virginia, served African American agricultural learners (Simmons, 1940).

These early agricultural clubs paved the way for the development of the Future Farmers of America. In 1928, the Future Farmers of America was established during the first National FFA Convention in Kansas City, Missouri. The national organization for the New Farmers of America (NFA) was created on August 4, 1935. Throughout the 1960s, the civil rights movement took a prominent position in our nation's history. As schools desegregated during this time, many FFA and NFA chapters merged at the local level. The passage of the Civil Rights Act in 1964 prompted both organizations to discuss the formation of one unified national organization (IUPUI University Library, 2016). On July 1, 1965, membership in the FFA was opened to all agriculture learners regardless of race, color, or national origin (Wakefield & Talbert, 2000). In 1969, after several failed attempts to amend the constitution, females finally gained the right to be members of the FFA (Moore, 2019).

SAE (Supervised Agricultural Experience)

It is thought that supervised experience was the earliest component of the agricultural education model as youth engaged in apprenticeships or received informal education at home (Croom, 2008). Rufus W. Stimson was credited with developing the project method of teaching in response to concerns about how agriculture was taught. He felt the instruction consisted of too much lecture and manual labor that did not actively engage learners. The project method proposed by Stimson allowed learners to transfer learning from the classroom to projects implemented at their home farms (Moore, 1988). The Smith-Hughes Act of 1917 proposed that learners should engage in directed or supervised practice in agriculture for at least six months during the year. Over time, this emphasis on meaningful and relevant applied learning has evolved into what it now known as Supervised Agricultural Experience (SAE). In 2017, the definition was updated to include that SAE is a "student-led, instructor supervised, work-based learning experience that results in measurable outcomes within a predefined, agreed upon set of Agriculture, Food and Natural Resources (AFNR) Technical Standards and Career Ready Practices aligned to a career plan of study" (The National Council for Agricultural Education, 2017, p. 2).

Types of Agricultural Education

In this textbook, we foreground agricultural education which occurs in a school setting (e.g., a high school agricultural classroom). This type of agricultural education is often termed "school-based agricultural education" or SBAE; however, there are other types of agricultural education. For example, county extension agents teaching community members improved farming practices is agricultural education. A director of an urban garden discussing pest management strategies with gardeners is also agricultural education. Agricultural education which occurs outside the scope of a school-centered program is referred to as "nonformal" or "community-based" agricultural education. Much like school-based agricultural education, agricultural education occurring outside the context of a school is critical to fostering learning and engagement in agriculture. While some of this textbook will focus exclusively on school-based, or "formal," agricultural education, *most* of the content will be applicable for those envisioning educating in either space.

Why Does Agricultural Education Exist?

Now that we have detailed the elements that compose agricultural education, we must situate *why* agricultural education exists. Agricultural education programs fall within the broader umbrella of Career and Technical Education (CTE). The objective of CTE programs is to prepare learners with the knowledge and skills needed for employment or further education in a corresponding field (e.g., business, health science, information technology). For agricultural education, exploration of and preparation for careers in agriculture, food, and natural resources are the foci. Our focus on vocational preparation has its roots in the Smith-Hughes Act of 1917, which established agricultural education as a vocational education program. Throughout its history, agricultural education has maintained an ethos of vocational preparation; however, other answers to the *why* of agricultural education have emerged, which we will consider in concert with the vocational aim.

Acknowledging the diversity of careers individuals enrolled in agricultural education courses have an interest in pursuing, agricultural education has evolved to develop *agricultural literacy*. Agricultural literacy entails knowledge and understanding within the discipline of agriculture (Clemons et al., 2018). Developing agricultural literacy among all learners, including those who do not plan to pursue agricultural careers, is important to create a society of citizens who are informed about issues that exist within agriculture, food, and natural resources. When possessing content literacy, citizens are more likely to make informed choices when voting, shopping, and communicating. Therefore, agricultural education exists to empower members of society with knowledge and understanding so they can improve their lives and the lives of others. Importantly, this literacy does not mean the role of agricultural education is to develop advocates for agriculture. Certainly, literacy often results in advocating; however, literacy can also yield critical thinking about how to continually improve agriculture for the present and future.

Agricultural education also exists to develop systems thinking among learners. Climate change, soil degradation, and water access are just a few of the complex problems which relate directly to agriculture, food, and natural resources. Agricultural education, therefore, must play a role in empowering individuals to understand and address these complex challenges (Pauley et al., 2019). Problem solving has long been a staple of agricultural education curriculum (Parr & Edwards, 2004). Building upon the traditions of problem-solving instruction, educators must seek opportunities to foster systems thinking and interdisciplinary mindsets through their curriculum. Systems thinking refers to identifying the multitude of relationships that exist among social, technological, and ecological factors within a given problem or phenomenon (McKim & McKendree, 2020). Systems thinking is imperative to identifying factors that influence a problem *and* factors influenced by potential solutions. Related to systems thinking, educators should seek to illuminate the content of multiple disciplines (e.g., science, technology, engineering, and mathematics [STEM]) within their instruction as these interdisciplinary learning experiences mirror the diverse mindsets needed to address complex problems. Building upon the need for interdisciplinary learning within agricultural education, there exists a timely opportunity for agricultural educators to highlight STEM career pathways for all learners. This opportunity is especially salient for populations underrepresented within STEM fields.

Leveraging the interdisciplinary nature of agriculture, many states and programs provide credit in core academic areas for learners engaging in agricultural education coursework. Offering agriculture courses which count for core academic credits positions agricultural education as a valuable option for learners to efficiently complete high school graduation requirements while reaping the myriad additional benefits of agricultural education programs. Further, some agricultural education programs have established articulation agreements with colleges and universities whereby experiences in agricultural education (e.g., individual courses, obtaining a state degree, and program completion) count for postsecondary credits, extending the value of agricultural education.

In addition to vocational preparation, content literacy development, problem solving, and credit obtainment, agricultural education exists to ignite learners' passion for continuing to learn. For many, the careers we envisioned as middle school and high school learners differs from what we actually end up doing. Therefore, agricultural education exists to "plant the seed"—to promote lifelong learning by illustrating the importance, complexity, and diversity of agriculture, food, and natural resource systems. As one's future unfolds, the passion sparked for agriculture during coursework could illuminate a new path within the disciplines of agriculture, food, and natural resources. While this answer to *why* agricultural education exists is the least clear, it can also be the most exciting! Engaging in agricultural education has the potential to impact learners far beyond their days of learning in high school, getting their first job, or picking a college major. The impact of agricultural education persists throughout one's life. Therefore, as agricultural educators we may not realize the extent of our impact on a life, community, or the world.

Who Does Agricultural Education Serve?

When looking around our classrooms on the first day of school, we are going to be surrounded by learners who all have different motivations for signing up for agricultural education. Some learners may have family members who are involved or employed in the agricultural industry. Their parents or grandparents might be farmers or ranchers, an aunt or uncle might work in an agriculture business, a cousin might be a food scientist, or a sibling might work in construction or metal fabrication. Another learner may have cultivated an interest in agriculture through personal interests or future career aspirations. It is common for learners who dream of one day becoming a veterinarian to enroll in an agriculture class. Likely, a couple of your learners will have no idea why they are in your classroom. The class may have just appeared on their schedule, or a guidance counselor may have encouraged them to take the course for a variety of reasons.

The content and delivery methods common to agricultural education might also attract learners to your classroom. Learners might find that the integration and reinforcement of numerous scientific concepts, as well as contextual teaching and learning, serves their needs and interests. The Hatch Act of 1887 provided the impetus for the first agriscience programs in the United States. As experiment stations were developed, district agricultural schools were also established to provide secondary instruction in agriculture and home economics. The schools delivered practical instruction in agriculture, while the experiment stations emphasized scientific application and research. Over time, agricultural education became more vocational in nature and focused more on training and less on academic instruction (Hillison, 1996).

However, in 1988, the need for the integration of scientific concepts in agriculture instruction was raised again by the National Research Council. Legislation, such as the Carl D. Perkins Vocational and Applied Technology Act of 1990, the School-to-Work Opportunities Act of 1994, and the Carl D. Perkins CTE Improvement Act, has repeatedly called for the integration of core content academics into Career and Technical Education (Threeton, 2007). Educators have embraced this opportunity to reinforce scientific concepts in their classroom and curricula has changed leading to course offerings such as agriculture biotechnology, food science, agriculture biology, agriculture chemistry, earth science in agriculture, and animal and plant sciences. In some states, several of these courses can be used to fulfill science requirements needed for admission to four-year institutions. Calculating feed rations, determining genetic probabilities, profit and loss margins, accurate measurements, and timber cruising are few topics in which learners can apply mathematical concepts Additionally, educators commonly use pedagogical strategies such as inquiry-based learning, experiential learning, and problem-based learning which support learners' scientific reasoning and contribute to student achievement in mathematics. Beyond the traditional academic content, learners can gain a lasting appreciation or even academic course credit in the visual arts when completing a unit or course in floral design.

Other learners may have had family members or friends who were in the FFA, so they may be counting down the days until they can have an FFA jacket of their own. Several factors have been identified as contributing to FFA participation such as encouragement from others, the positive impact of teachers on their learners, and opportunities to develop professional and personal skills that look favorable on college applications or résumés. Learners also seek participation in the social aspects of the FFA organization and engage in activities beyond the classroom, such as field trips and community engagement (Phelps et al., 2012).

Even with numerous opportunities and benefits associated with the FFA, some learners may be reluctant to become members. As agriculture teachers look to support and offer options for all learners, they must also be cognizant of the student population they may not be serving. Learners might have negative perceptions or misconceptions about learners involved in the organization or may just be apathetic toward involvement in extracurriculars. There could also be challenges with schedules, both during school and after school. Learners may not have time to include an agriculture class in their class schedule or might feel like they do not have any extra time outside of school to participate in FFA events (Phelps et al., 2012).

How is Agricultural Education Delivered?

School-based agricultural education comprises three interrelated components: classroom and laboratory instruction; experiences in Career and Technical Student Organizations such as the National FFA Organization (FFA), the National Post-Secondary Agricultural Student Organization (PAS), or the National Young Farmer Educational Association; and experiential-, service-, and/or work-based learning through the initiation and continuation of a Supervised Agricultural Experience (SAE) program. The classroom and laboratory component consists of the planning and delivery of instruction and assessment of student learning. While content knowledge remains consistent across the country, there is variation in the curriculum guides/blueprints/frameworks, instructional materials and resources, and tests and performance assessments that are used in agricultural education.

The goal of Supervised Agricultural Experience (SAE) is to provide learners individualized opportunities to apply principles in the classroom and laboratory to a real-world context. SAE can also promote career exploration, support the acquisition of technical skills, and develop dispositions contributing to the college and career readiness of learners (The National Council for Agricultural Education, 2017). When starting SAE programs, learners are encouraged to complete a Foundational SAE. This allows learners to discover the broad scope of the agricultural industry, develop personalized interests in agriculture, and identify possible career opportunities. From the completion of a Foundational SAE, learners can expand their SAE with an Immersion SAE. There are five types of Immersion SAEs: Placement/Internship, Ownership/Entrepreneurship, Research, School-Based Enterprise, and Service Learning. Financial literacy, workplace safety, and key employability skills are also reinforced throughout the completion of SAE (The National Council for Agricultural Education, 2017). Agricultural educators play an important role by helping their learners develop plans and set goals, providing ongoing supervision, encouraging continual reflection and growth, and incorporating and reinforcing record keeping skills.

The FFA is an intracurricular student organization that complements both classroom instruction and SAE (Croom, 2008) and offers a variety of career and leadership development events, leadership conferences, award programs, and scholarships. The numerous opportunities allow for the development of key leadership skills and encourage personal growth among members. The National FFA Organization also promotes the importance of service engagement in local communities through community service and service-learning projects. Additionally, the opportunity to travel, both domestically and internationally, can provide FFA members a chance to experience new places, differing perspectives, and other cultures. For example, FFA members in North Carolina were able to host Future Farmers of Japan members from Tokyo Metropolitan Engei High School. A FFA chapter in Ohio traveled to Brazil and the Dominican Republic. The North Huron FFA in Michigan sends a few members to Uganda each year as an outreach opportunity and personal growth experience. While travel may not always be feasible, FFA chapters can also connect and engage through various social media outlets and video conferencing.

Numerous groups and entities contribute to the effective delivery of agricultural education. In 1983, the National Council for Agricultural Education was established to provide leadership and identified important opportunities and needed initiatives to support the ongoing development and direction for agricultural education (The National Council for Agricultural Education, 2012). Additionally, at the national level and throughout many states, Team Ag Ed provides a collaborative effort among secondary and postsecondary educators to improve student achievement. At the national level, organizations contributing to Team Ag Ed include the National Council for Agricultural Education, US Department of Education, National Association of Agricultural Educators (NAAE), National Association of Supervisors of Agricultural Education (NASAE), American Association for Agricultural Education (AAAE), Association for Career and Technical Education, National Farm and Ranch Business Management Education Association, Inc., National Young Farmer Education Association, and National Professional Agricultural Student Organization (PAS). Individual states may create their own Team Ag with state-specific representatives along with participation from additional groups such as Farm Bureau, the Department of Agriculture, FFA Alumni Association, FFA Foundation, and the agricultural industry.

Learning Confirmation

As we presented, the aim of this chapter was to introduce you to agricultural education. As we wrap up the chapter, we want to challenge you to take the content you have learned and create a personal vision statement for your career in agricultural education. Within this vision statement, identify *what* your career in agricultural education will look like; specifically, describe your vision for facilitating instruction, FFA, and SAE. In addition, address (*a*) *why* your work in agricultural education will be important, (*b*) *who* you envision educating throughout your career, and (*c*) *how* you envision delivering your agricultural education program. To prepare you for writing this vision statement, we recommend you take a moment to first complete the ten reflection questions provided at the end of this chapter. Finally, because this vision statement is personal to you, we recommend crafting the vision statement in a format that aligns to you as an individual. Example formats could include a video, a written paper, a podcast, personal website, or a picture or painting. Once you have crafted your vision statement, we encourage you to keep a copy and continually revisit and revise your vision throughout your experiences in agricultural education.

Applying the Content

For additional applications of the content, please consider the following:

1. Using the "Setting the Stage" section as an example, write out what you want a student to say during your retirement celebration.
2. Interview three current agricultural educators to identify *why* they teach, *who* they teach, and *how* they deliver their agricultural education program. Compare and contrast the information you receive with peers conducting similar interviews.
3. Conduct interviews with students in agricultural education at your home institution to identify their motivation(s) for being in agricultural education and their vision for their careers as agricultural educators. Compare the data you collect to your own motivations and vision.

Reflective Questions

After reading this chapter, please reflect on the following questions about our futures as agricultural educators:

1. Reflect on effective agricultural educators we have engaged with; what attributes make them effective?
2. What types of instructional resources (e.g., laboratories, technology, barns) do we want to have available in our future programs?
3. What types of SAE opportunities would we like to encourage our future learners to pursue?
4. What are traditional FFA experiences we would like to include in our future program and what are some unique or new FFA opportunities we would like to provide for our future learners?
5. What are five goals we would like our future FFA program to achieve?
6. How are we going to promote diversity, equity, and inclusion within our future program?
7. After exploring the historical development of agricultural education, what might be some future changes we anticipate within the discipline?
8. What are some ways we would like our future agricultural education programs to support community vitality?
9. What external organizations would we like to leverage to support our professional growth as agricultural educators?
10. In looking forward to reading the remainder of this textbook, what are we most interested in learning about?

Glossary of Terms

- **agricultural literacy:** Knowledge and understanding within the discipline of agriculture.
- **career and technical education:** Secondary and postsecondary coursework providing students with academic and technical skills, knowledge and training necessary to succeed in future careers and become lifelong learners.
- **community-based agricultural education:** Agricultural education which occurs outside the scope of a school-centered program.
- **National FFA Organization (FFA):** A youth leadership organization that strives to make a positive difference in the lives of young people by developing their potential for premier leadership, personal growth, and career success through agriculture education
- **Supervised Agricultural Experience (SAE):** a planned and supervised program of experience-based learning activities that extend school-based instruction and enhance knowledge, skills, and awareness in agriculture and natural resources
- **instruction:** Purposeful direction of the learning process through the use of teaching methods and strategies.
- **school-based agricultural education (SBAE):** Systematic program of instruction in agriculture, food, and natural resources for K-12 students within a school program.
- **vocational development:** Developing the skills and knowledge to support future employment.

References

Clemons, C., Lindner, J. R., Murray, B., Cook, M., Sams, B., & Williams, G. (2018). Spanning the gap: The confluence of agricultural literacy and being agriculturally literate. *Journal of Agricultural Education*, *59*(4), 238-252. https://doi.org/10.5032/jae.2018.04238

Croom, D. B. (2007). Agricultural education at the Tuskegee Normal and Industrial School. *Journal of Agricultural Education*, *48*(2), 13–22. https://doi.org/10.5032/jae.2007.02013

Croom, D. B. (2008). The development of the integrated three-component model of agricultural education. *Journal of Agricultural Education*, *49*(1), 110–120. https://doi.org/10.5032/jae.2008.01110

Davis, B. M. (1912). *Agricultural education in the public schools*. University of Chicago.

DeLay, A. M., & Swan, B. G. (2014). Student apathy as defined by secondary agricultural education students. *Journal of Agricultural Education*, *55*(1), 106–119. https://doi.org/ 10.5032/jae.2014.01106

Engelhardt, E. (2009). Canning tomatoes, growing "better and more perfect women": The girls' tomato club movement. *Southern Cultures*, *15*(4), 78–92. https://doi.org/10.1353/scu.0.0087

Evans, L. B., Duncan, L. N. & Duncan, G. W. (1916). *Farm life readers, book* 5. Silver, Burdett & Company.

Glatthorn, A. A., Carr, J. F., & Harris, D. E. (2001). *Curriculum handbook: Planning and organizing for curriculum renewal.* Association for Supervision and Curriculum Development.

Hillison, J. (1993). The Role of Virginia in the development of the FFA. *Journal of Agricultural Education 34*(2), 37–45. https://doi.org/10.5032/jae.1993.02037

Hillison, J. (1996). The origins of agriscience: Or where did all that scientific agriculture come from? *Journal of Agricultural Education, 37*(4), 8–13. https://doi.org/10.5032/jae.1996.04008

IUPUI University Library (2016). New Farmers of America Records, 1929-1965 [Online Archives]. Ruth Lilly Special Collections and Archives, Indiana University Purdue University. https://archives.iu.edu/catalog/mss059

McKim, A. J., Velez, J. J., Lambert, M. D., & Balschweid, M. A. (2017). A philosophical review of science and society within agricultural education. *Journal of Agricultural Education, 58*(2), 98–110. https://doi.org/10.5032/jae.2017.02098

McKim, A. J., & McKendree, R. B. (2020). Metacognition, systems thinking, and problem-solving ability in school-based agriculture, food, and natural resources education. *Advancements in Agricultural Development, 1*(1), 38–47. https://doi.org/10.37433/aad.v1i1.21

Moore, G. E. (1988). The forgotten leader in agricultural education: Rufus W. Stimson. *Journal of the American Association of Teacher Educators in Agriculture, 29*(3), 50–58. https://doi.org/10.5032/jaatea.1988.03050

Moore, G. E. (2019, January 25). The age of aquarius: The 1960s and girls in the FFA. *The Friday Footnote.* https://footnote.wordpress.ncsu.edu/2019/01/24/the-age-of-aquarius-the-1960s-and-girls-in-the-ffa-1-25-2019/

The National Council for Agricultural Education. (2012). *About us.* https://thecouncil.ffa.org/about-us/

The National Council for Agricultural Education. (2017). SAE for all teachers guide. https://ffa.app.box.com/s/exollg1x7q2lntun3su2mdufw07wiklf

National Research Council. (1988). *Understanding agriculture: New directions for education.* Washington, DC.

North Carolina Career and Technical Education. (2020). *Updated CTE course blueprint: Sustainable agriculture production I.* https://ncffa.org/agricultural-education/curriculum/

Parr, B., & Edwards, M. C. (2004). Inquiry-based instruction in secondary agricultural education: Problem-solving – an old friend revisited. *Journal of Agricultural Education, 45*(4), 106–117. https://doi.org/10.5032/jae.2004.04106

Pauley, C. M., McKim, A. J., & Hodbod, J. (2019). A social-ecological resilience perspective for the social sciences of agriculture, food, and natural resources. *Journal of Agricultural Education, 60*(4), 132–148. https://doi.org/10.5032/jae.2019.04132

Phelps, K., Henry, A. L., & Bird, W. A. (2012). Factors influencing or discouraging secondary school students' FFA participation. *Journal of Agricultural Education*, *53*(2), 70–86. https://doi.org/10.5032/jae.2012.02070

Pound, K., & Moore, G. (2012, February 4-7). *Achieving equity for females in agricultural education: The development of tomato clubs* [Paper presentation]. Southern Region AAAE 62nd Annual Meeting, Birmingham, AL, United States.

Reck, F. M. (1951). *The 4-H story*. The Iowa State College Press.

Simmons, S. B. (1940). Negro youth and the U. S. junior employment service - 4-H clubs and the New Farmers of America. *The Journal of Negro Education*, *9*(3), 408–415.

Threeton, M. D. (2007). The Carl D. Perkins Career and Technical Education (CTE) Act of 2006 and the roles and responsibilities of CTE teachers and faculty members. *At Issue*, *44*(1), 66–82.

True, A. C. (1929). *A history of agricultural education in the United States 1785 - 1925*. United States Department of Agriculture Miscellaneous Publication No. 36. Government Printing Office.

Wakefield, D. B., & Talbert, B. A. (2000, December 6). *Exploring the past of the new farmers of America (NFA): The merger with the FFA* [Paper presentation]. National Agricultural Education 27th Annual Research Conference, San Diego, CA, United States.

Yeatts, A. L., Jr. (1954). *A history of the Future Farmers of America in Virginia* [Unpublished master's thesis]. Virginia Polytechnic Institute and State University. http://hdl.handle.net/10919/41197

2. PSYCHOLOGY OF LEARNING

Hui-Hui Wang and Summer Odom

Setting the Stage

As your first year of teaching draws to a close, you find yourself reflecting on the year. You were hired as one of three agriculture teachers to teach floral design, vet science, and welding. Of the 1,500 students in the school, you begin to think about the students who walked into your class each day. While a majority achieved a passing grade, what did they really learn in your courses? As you consider the makeup of your classes, you realize that one in five students were from underrepresented groups and two out of five students did not have an agricultural background. You question whether you have adequately met the needs of these students especially as they were not the ones who won awards or competed in career development or judging events. It leaves you wondering if they learned what you expected them to learn in your courses. You begin to reflect on several important questions: Are there changes you could make to your instructional approach to better excite all your students, regardless of their background? Is there research about learning to help you design strategies to ensure your teaching results in meaningful learning outcomes for all students?

Objectives

This chapter serves to highlight what we know about how learning occurs including some of the theories that have been proposed by scholars about the principles and processes that influence learning. There are a vast amount of learning theories and we will not attempt to review all theories in this chapter (for a more complete list, see Schunk, 2012). Instead, we will focus on those theories that are applied versus basic theories of learning and most applicable to educators throughout agricultural education programs. Furthermore, we will review theories that help situate the environment of learning, including the identities of learners and diversity and inclusion aspects of learning. It is our intent that this chapter can be used to help explore beliefs and assumptions about learning that will help guide teaching efforts in the classroom and beyond and enable you to align your instructional strategies to create a dynamic learning environment for your learners.

- Discuss the definition of learning and overview of foundational learning theories.
- Describe specific learning theories that have implications for building dynamic teaching practices for high school agriculture programs.
- Discuss how identities of learners impact their learning in the classroom including diversity and inclusion aspects in learning.
- Recognize ways to apply theories and frameworks of learning in your classroom.

Introduction

Research on learning should ultimately contribute to improved teaching. While teaching and learning complement one another, good teaching and good learning are not sufficient to ensure one another's success. Educators play a critical role in creating effective learning environments that assist in completing cognitive activities necessary to develop and demonstrate certain skills and abilities (Schunk, 2012). But your actions may not always result in learning by your participants. Many factors influence whether learning occurs or not, and thus it is important to have a deeper understanding of what learning is and how it happens. So, let's go a little deeper into a definition of learning to help better understand how learning happens. Does learning involve an increase in knowledge? Does learning involve a change in behavior? Does learning require reflection? Does learning require that you demonstrate some type of activity? Does learning involve a change in beliefs and attitudes? Does learning happen better through social interactions? While researchers, practitioners, and theorists have not reached an agreement on a universal definition of learning (Schuell, 1986), Schunck (2012) defines learning as "an enduring change in behavior, or in the capacity to behave in a given fashion, which results from practice or other forms of experience" (p. 3). Three criteria for learning include: (1) learning involves change, (2) learning endures over time, and (3) learning occurs through experience.

Just as there is no universal definition of learning, there is also debate on how learning occurs (processes) and factors that impact learning. Older theories of learning center on behavioral views, meaning learning is explained through observable behaviors. For instance, if you were an educator who believed learning occurred this way, you would arrange the environment so that individuals would respond to certain features. For example, you give a time-out to learners who need to restore their attention to your class. A cognitive approach situates learning as an internal mental phenomenon that you can infer from what people say and do. An implication for teachers as a result of cognitive theories would mean teachers should make learning meaningful and take into account learners' perceptions of themselves (their identity) and their environment. The constructivist perspective of learning contends that individuals learn by constructing or forming their knowledge. It focuses more on human factors that explain how learning occurs. Most modern learning theories are based on the tenets of cognitive and constructivist theories so in this chapter, our focus will be on the cognitive and constructivist aspects of learning.

Though many of the learning theories differ in how they distinguish the processes and factors of learning, we know learning is not just about the instructional factors (Pintrich et al., 1986). Of critical importance is what learners do with the information and how they process this information (Schunk, 2012). We also know educators cannot ignore the differences among individual learners including their thoughts, beliefs, attitudes, and values. These factors influence the learning process.

Motivation impacts how individuals learn. In this chapter, we will also explore some theories that offer explanations for what motivates learners. How people approach and construct learning is impacted by their identities (e.g., perceptions, values, goals, beliefs, and attitudes). The identity exploration framework offers a model to consider as we think about how to engage and motivate learners.

We will introduce some theoretical and scholarly foundations of learning and the brain, behavioral, cognitive, and constructivist theories of learning and the tenets of motivation related to learning. Specifically, we will explore the following content related to learning:

1. Learning and the brain
2. Overview of learning theories: Behaviorism, cognitivism, and constructivism
3. Selected learning theories: Information processing theory and cognitive load theory
4. Selected motivation theories: Social cognitive theory, outcome expectations and goal setting theory, self-efficacy, and identity exploration framework
5. Synthesis of learning theories and research related to learning

Learning and the Brain

How does understanding the function and structure of the human brain increase our understanding of being an effective educator? Understanding brain development and the location of brain functions helps constrain educational theories and models of behavior (Casey et al., 2000). Human brain development is a complex event occurring across the life cycle. Through the discoveries of cognitive neuroscience, we learned that humans receiving, selecting, storing, transforming, developing, and recovering information are all parts of a continuous development process that happens from birth to early adolescence (Casey et al., 2005). Human brain development in the temporal and frontal areas have been consistently associated with higher cognitive abilities (Casey et al., 2005, 2002; Krawczyk, 2018). Research in neuroscientific brain maturation of cognitive abilities, such as short-term (or working) memory and long-term memory, provides valuable information that can be applied in teaching and learning. Some distinct findings from cognitive neuroscience include: (1) Short-term memory has limited capacity and is associated with the focus of attention (Cowan, 1995); (2) Sensory functions, such as visual, short-term memory capacity declines with increasing task complexity (Alvarez & Cavanagh, 2004); (3) Instead of cramming information in a long and intense learning session, repeating information in multiple short learning sessions has optimized long-term memory and improved information retention (Gluckman et al., 2014); (4) Different kinds of memories rely on different neural correlates and mental processes (Nie et al., 2019), (5) Connecting prior knowledge with new information promotes comprehension and memory coding performance (Maguire et al., 1999), and (6) Various instructional approaches stimulate different parts and functions of the human brain that impact short-term and long-term memory—for example, comparing the repeated practice of solving a mathematics equation to letting learners create a solution stimulates different patterns of brain activity that help learners use more short-term memory (Wirebring et al., 2015).

Although understanding the function and structure of the human brain shines light on how people learn, learning is a far more complicated process than simply understanding neuroscientific brain maturation of cognitive abilities. A more recent and dominant learning perspective in the discipline of learning sciences is the 4E Cognition model. 4E Cognition contends that learning is enacted, embodied, embedded, and extended (Steier et al., 2019). 4E Cognition is a tool for understanding the complicated notion of cognition (or learning). The premise of the 4E Cognition model is that learning involves more than just the brain; our bodies, the situation, and interactions within the environment also contribute to learning. The enacted part of the model implies that thinking involves interaction with others, the immediate context, and various other items like language and culture. Embodied implies that learning is in relation to our actions. Our actions imply learning. The embodied mind cannot function in isolation from the physical, social, and cultural environment because the environment in which the mind exists is part of the mind. Embedded has to do with the environment including physical elements, cultural-historical structures, and social characteristics like interpersonal relationships and interactions. The unique features of an environment provide unique affordances for cognition. For learning to be effective, environmental resources should be taken into account (Pouw et al., 2014). To help you think about the idea of 4E Cognition, consider this example. Think about the context of baseball and specifically an outfielder catching a fly ball. Before the batter steps to the plate, the coach might be able to speculate where the outfielder should stand based on past performance and experience with that particular batter. But, when it comes time for the outfielder to catch the ball, they must react in the moment and adjust according to where the ball is actually hit.

Overview of Learning Theories

Behaviorism, cognitivism, and constructivism play an important role in the development of contemporary educational psychology theories. In the early twentieth century, behaviorism emphasized using instructional reinforcement (both positive and negative) to structure a learning environment, and behaviorists evaluated observable behavior and performance to measure success of learning (Skinner, 1976). Although receiving many criticisms, behaviorism has proven as an effective instructional strategy when a teacher uses memorized educational materials and instructions (Woollard, 2010). For example, when an educator gives immediate and direct feedback to praise learners who correctly answered a pop quiz (factual questions), it is generally recognized as a behaviorist's teaching strategy. In the 1950s and 1960s, there was a shift from behaviorism to cognitivism, where instructional design was focused more on thought processes and mental activities but less on changing behaviors (Wittrock, 1986).

Cognitive psychologists believe that learning occurs through internal mental processing of information and that learners' thoughts, beliefs, and attitude play important roles in their learning process (Winne, 1985). Cognitivism stresses that learning is an active and goal-oriented process (Shuell, 1986), and cognitivism utilizes external tools as an instructional strategy to provoke learners' internal mental activities to learn efficiently. For example, when an educator asks learners to keep an agriscience journal to reflect on what they have learned, this is a cognitive learning strategy that helps learners organize their learning based on how the human brain processes information internally.

On the other hand, constructivism is a learning theory that highlights how learners construct knowledge by developing personal meaning out of experiences rather than passively receiving information. Learning involves the reconstruction of existing knowledge to include the newly encountered experience, and existing knowledge has a contextualized meaning that is aligned with the experience (Haglud et al., 2012; Piaget, 1970). In a classroom, constructivists believe an educator needs to help learners retrieve their existing knowledge and provide opportunities for learners to reconstruct their existing knowledge with external experiences. Additionally, from the social constructivist viewpoints, social interactions play an important role in learning (Vygotsky, 1978). Learners construct their knowledge through conversation, discussion, and interaction with their peers and with educators. For example, instead of asking factual questions, an educator asks "why" questions to elicit learners' prior knowledge and continually facilitates conversations with learners. The educator attempts to structure a learning environment that allows learners of different backgrounds and experiences to share their thoughts and construct knowledge depending upon their needs. Table 2.1 summarizes the three dominant paradigms of learning theories, behaviorism, cognitivism, and constructivism.

Learning Theory	Overall Definition of Theory	Implications for Classroom Strategies
Behaviorism	Learning involves behavior changes and can be studied in a systematic and observable way.	Giving immediate and direct feedback; Giving learners a "time-out" or detention; Giving learners awards and prizes.
Cognitivism	Learning involves internal mental processing of information. Learning is an active and goal-oriented process.	Using flashcards to remember vocabularies; Taking notes; Writing a summary; Using a concept map to organize information.
Constructivism	Learning involves social interactions and the reconstruction of existing knowledge to include the newly encountered experience.	Using KWL (what I know, what I want to know, what I learned) instruction; Using inquiry-based instruction; Having learners do group discussions and reflections.

Table 2.1: Summary of behaviorism, cognitivism, and constructivism.

There are many educational psychology theories that are rooted in behaviorism, cognitivism, and constructivism. In this chapter, we go particularly deep into information processing theory (an extension of cognitivism) and some theories that are associated with motivation and learning, such as social cognitive theory and the identity exploration framework.

Selected Learning Theories

Information Processing Theory

Why do learners recall some information but not other information? Information processing theory (IPT) explains how the brain works to learn, store, and use information. The theory attempts to explain the path of how information is encoded into the memory system. Some scholars use an analogy to compare IPT with computer coding and analyzing information. Four factors—stimulation or stimuli, sensory registers, short-term memory, and long-term memory—are involved in the IPT. When we receive environmental stimuli through our senses (sensory input), we try to make meaning of them. In human brains, each sense has its own register to make meaning of the sensory input. Sensory information catches our attention, and the sensory registers progress the information to the short-term memory. However, only the information that is deemed relevant or is familiar to previously stored information will be transferred from sensory registers to short-term memory. The screening process is very selective. In other words, if there are many stimuli, we only respond to certain stimuli that catch our attention. Unless novel information is transferred from short-term memory to long-term memory, learning has not occurred. Our brains need to encode (elaborating processes in transferring) the information before we can store them into long-term memory. During encoding, our brains try to integrate the new information with the knowledge already stored in long-term memory and try to organize the information into a meaningful way. Retrieving existing knowledge and integrating newly encountered information with existing knowledge is critical in helping information transfer from short-term memory to long-term memory. Piaget (1970) provided a different perspective through the lens of educational psychology to explain this process. Piaget suggested two ways individuals use to construct knowledge: *accommodation*, in which individuals try to develop a new schema (as building blocks of our memories to help us understand how things work) to cope with a new experience, or *assimilation*, in which individuals try to balance and adapt the existing schema to a new experience. When experiences do not fit into an existing schema, individuals seek *equilibration*. Individuals take actions, such as asking questions, repeating the new experience, or seeking more information, to find equilibration to make sense of the newly encountered experience (Piaget, 1970).

Cognitive Load Theory

Another widely accepted learning theory based on human brain information processing and storing is cognitive load theory (Sweller et al., 1998). Cognitive load theory contends given that working memory is limited to processing novel information, working memory load is affected by the complexity of the task (intrinsic cognitive load), the relationship between the learning goal and design of instructional materials (extraneous cognitive load), and degree of integration of new information with prior knowledge that is stored in the long-term memory (germane cognitive load) (Schunk, 2012; Sweller, 1994). Therefore, when designing instruction, an educator should remember not to "overload" students with too much new or unfamiliar material, teach in a context with minimized distractions, and break complex learning into smaller chunks to not overwhelm cognitive load.

Motivation and Learning

Motivation is the study of why individuals behave the way they do. Educators are often challenged to figure out how to motivate their learners and interest them in being engaged in their learning and completion of tasks in a learning environment, such as a classroom or laboratory. While there are still many unanswered questions about motivation and little rigorous and systematic research on the interventions of the constructs of motivation, there are some tenets of motivation that are useful to apply in in formal and nonformal learning environments (Graham & Weiner, 2012). Motivation is a driving force that leads to goal-directed behavior to achieve an objective or a certain level of performance. Educators are driven by increasing learners' motivation and their excitement to learn. What inspires people to learn and achieve the educational goals set for them? How can educators promote learners' engagement and motivation to learn?

Broadly speaking, there are two types of motivation: intrinsic and extrinsic. Intrinsic motivation refers to when a person is moved to act for their interests, satisfaction, and enjoyment. Extrinsic motivation means a person is moved to act for external rewards, such as money, praise, or to avoid punishments. In this section, we introduce social cognitive theory, outcome expectations and goal setting theory, self-efficacy, and identity exploration framework and provide examples of how these theories could be used to promote learners' motivation and performance.

Social Cognitive Theory

Social cognitive theory (SCT) is one of the landmark theories developed by psychologist Albert Bandura (1986). The theory stresses dynamic and reciprocal influences of behavior, environment, and individual cognitive factors. Learning through observation, outcome expectations, and self-efficacy are considered as the three individual cognitive factors that are affected by the environment to shape individuals' behavior. Learning can occur by observing and watching others, which also is known as modeling, imitating, or mirroring (Meltzoff, 1990). Studies have shown observational learning involves both behavioral and neurophysiological reactions (e.g, mirror neuron system) to social environments (Csibra, 2007; Rizzolatti, 2005). Observers take note of others' behavior, comprehend the meaning of the behavior, and imitate the behavior. For example, children might pick up on their parents being afraid of bugs, and then they also exhibit fear of insects. Specific to a learning environment, apprenticeship is one of the most representative samples of learning through observation. Apprenticeship, where an experienced master-mentor trains the newcomers to acquire skills and knowledge, is a common and effective way that health care, manufacturing, construction, and engineering are used to structure on-the-job training or classroom learning. In education, student teaching and early field observation are common practices of apprenticeship.

Outcome Expectations and Goal Setting Theory

Outcome expectation is one of the critical factors in SCT. Outcome expectation is defined as anticipated consequences (positive or negative) of a person's behavior (Bandura, 1986). Outcome expectations highly connect with setting a goal and taking actions toward behavior change. Having learners set goals for themselves is one of the most effective ways to stay motivated. Locke and Latham (1990, 2002) proposed five principles of effective goal setting: clarity, challenge, commitment, feedback, and task complexity. Setting up a specific, clear, and challenging goal is better than a general, vague, and easy goal. For example, instead of setting up a goal like "I want to improve my overall course grade," a specific and clear goal could be "I want to improve my overall course grade from C to B." In the goal setting process, a learner should also put deliberate effort into achieving this goal. For example, "I want to improve my course grade from C to B. Especially, I want to increase crop and weed science by fifty points. To meet this goal, I will study the textbook for two hours a week." Receiving immediate feedback is a key component for goal setting. An educator's role is providing regular feedback throughout the process and helping learners to keep track of their progress. Additionally, if the goal is too complex, an educator could help break down the process into subgoals or give feedback on setting up a realistic timeline. For example, an educator's feedback could be "I suggest you also put more details about what you will study in the textbook. For example, in week one and week two, you could focus on genetics. Then, in week three and four, you can move onto studying plant breeding and biotechnology." Goal setting theory is highly associated with self-efficacy.

Self-efficacy

Bandura discusses the essential role that self-efficacy plays in social learning theory (Bandura, 1977, 1986). Self-efficacy refers to an individual's belief in their own capacity to accomplish goals, which influences academic motivation, learning, and achievement (Bandura, 1986). Self-efficacy is task specific, and it is a strong predictor of task performance (Wood & Bandura, 1989). Learners who are more efficacious may have higher expectations for themselves and will set up higher level goals than those learners with less efficacy. Learners who are more efficacious are also more committed to achieving goals and respond more positively to failures (Locke & Latham, 1990). Bandura (1977) suggests four sources to increase self-efficacy: mastery experiences, vicarious experiences, verbal feedback and persuasion, and emotional and physiological states. To help learners master their learning, an educator can create a positive learning environment through reinforcing positive behaviors, using high-energy interactive instruction, and nurturing positive relationships to help learners value what they do. Educators can help learners know that although challenges and failures are inevitable, they have abilities, confidence, and desire to achieve their goals. Educators should also seek out various opportunities for learners to see successful role models. When seeing people with similar backgrounds, knowledge, and skills have successful experiences, learners increase their confidence and belief that they will experience similar success. When educators provide positive feedback to learners, it can lead to higher perceptions of self-efficacy and enhance learner success. Additionally, the emotional and psychological well-being of a person can influence how they feel about their capacities in a particular task.

Identity Exploration Framework

Much of the recent motivational research is focused on awareness and acknowledgment of individual differences and categorization of individuals as being either mastery or performance oriented, intrinsically or extrinsically motivated, and high or low in self-efficacy and self-regulation. In regard to the individuality of learners, educators must recognize that individual attributes of learners are important to how they learn (Graham & Weiner, 2012). Researchers, policy makers, teachers, parents, and learners themselves recognize that developing personal and interpersonal attributes in today's changing world requires educators to focus on promoting learners' agency and figuring out who they are and who they will become, the formation of their identities. Identity exploration has been defined as "deliberate internal or external action of seeking and processing information in relation to the self [where the outcome is], the creation of self-relevant meaning with an integrative effect and the facilitation of development" (Flum & Kaplan, 2006, p. 100). It is recommended that an educator's role in identity exploration "is to organize academic experiences and opportunities that would encourage learners to question their self-aspects and investigate and consider alternative perceptions, values and goals" (Sinai et al., 2012, p. 197). Examples of pedagogical practices educators can engage in to promote identity exploration include posing complex personal problems that encourage learners to think about strengths and limits to alternative solutions, exposing learners to others who are experiencing identity exploration, using writing activities that encourage self-reflection, and creating focused activities around decision-making and critical thinking in the curriculum (Waterman, 1989).

The Identity Exploration Framework (Flum & Kaplan, 2006) is a central mechanism for identity formation associated with intense engagement, positive coping, openness to change, flexible cognition, and meaningful learning. Identity is considered an individual's perceptions, values, beliefs, attitudes, and goals, is crucial to becoming, and is focused on adaptive motivation for learning the academic material. Some principles that have been identified for promoting adaptive identity exploration within the school curriculum (Kaplan et al., 2014) include:

1. Promoting relevance
2. Triggering exploration
3. Facilitating a sense of safety
4. Scaffolding exploratory actions

Promoting relevance. Promoting relevance involves helping learners connect academic content to self-aspects such as abilities, attributes, goals, values, behaviors, and emotions. It is important to note that perceived relevance is subjective in that it depends on the learner's current self-constructions, concerns, interests, background, and experiences. One strategy for promoting relevance is to ask the learners to make connections between the academic content and aspects of their lives. For example, as a high school agriculture educator, one place where promoting relevance to learners may be natural is through opportunities like Future Farmers of America (FFA) and participating in Supervised Agricultural Experiences (SAE) events. This is where learners' individual identities and interests could be illuminated. This strategy can be useful in increasing success of learners by triggering exploration, facilitating a sense of safety, and scaffolding exploratory actions.

Triggering exploration. Triggering exploration involves the questioning and examining of self-aspects that become apparent to learners through relevance. Exploration triggers are subjective experiences that are different from the current identifications or identity commitments (self-perceptions, values, goals, social roles, and relationships) of learners. Because learners differ in their concerns, cultural backgrounds, and openness to identity exploration, the same event may be an exploration trigger for one learner but not for another. Designing experiences that trigger exploration are reliant on the educator's familiarity with the learners. Exploration triggers may also provoke a sense of conflict and threat including threats to self-worth that lead to struggle and resistance. Educators should be mindful of ensuring a safe learning environment to reduce threats associated with exploration triggers.

Facilitating a sense of safety. Exploration trigger experiences that are perceived as threats, are disappointments to significant others, or are at odds with one's internalized sense of self can lead to learners feeling anxious and defensive and may require educators to establish a sense of safety in order for learners to engage in exploration. Like the previous two principles, learners may differ in what they perceive as being safe. Educators should engage in strategies that promote belonging, establish norms of mutual respect, and legitimize the perspectives and emotions of learners.

Scaffolding exploratory actions. Once learners perceive relevance of academic content to the self, experience a "relevant different" to trigger their identity exploration, and feel safe to engage in this exploration, they still need to have the knowledge in effective exploration strategies. Examples of scaffolding identity exploration activities include reflective questions for personal and interpersonal consideration, role-playing exercises, reflective writing activities, and peer modeling.

Educators should recognize the unique aspects of learners, especially in regard to the multicultural aspects of learners. Educators should continue to assess the psychological state of learners as they design activities: Are the learners perceiving self-relevance of the content? Are learners experiencing a "relevant difference" from identity aspects? Do learners feel safe? Are learners engaging in adaptive exploration strategies? Reviewing answers to these questions can assist educators in deciding which actions they should take to further learners' identity exploration in relation to the academic material. Likewise, it is suggested that educators may not be prepared to engage in identity exploration because they do not hold this approach to teaching. Educators may ask themselves the same questions, but with the emphasis being on themselves instead of the learner (Kaplan et al., 2014).

In summary, there are many theories and research on learning that attempt to explain how learning occurs. While many theories exist (for a more comprehensive list, see Schunk, 2012), we have described some of the theories most relevant to agricultural education. The three dominant paradigms of learning theories are behaviorism, cognitivism, and constructivism. Within these paradigms are more specific theories like social cognitive theory. Motivation and self-efficacy are important factors for learning and should be accounted for when designing programs that maximize student learning in your classroom. Finally, the identity exploration framework is a model for educators to use in helping learners with identity exploration since identity and individual attributes of learners has been found to affect learning (Graham & Weiner, 2012).

Learning Confirmation

1. Which learning theory is being demonstrated when you give learners awards for completing tasks?
2. How could you increase an individual's motivation to learn by applying goal setting theory?
3. How can you create relevance for learners?
4. What implications for learning does the identity exploration framework present for working with learners from underrepresented populations?
5. How could you increase an individual's self-efficacy in your program?

Applying the Content

In this chapter, we have attempted to provide an overview of several theories of learning most relevant to agricultural education. While you may not always think about the specific theories as you teach, these theories and research on learning can be applied and are already being applied in pragmatic ways throughout an agricultural education program. For example, with the three-circle model used in high school and middle school agricultural programs, learners have more advantages than traditional learners to "learn" (chapter 3) through being in the FFA program (chapter 10) and through their SAEs (chapter 11). We want to highlight a few applications of learning theory (experiential learning, discovery learning, and project-based learning) that are prevalent within agricultural programs.

Experiential learning has been recognized as a critical component of a comprehensive agricultural education (Baker et al., 2012). As an application of constructivist learning theory, experiential learning involves learners experiencing something and then reflecting on this experience to make meaning of the experience. It is important to note that for learning to occur, agricultural educators must be present and purposeful through all program components and ask reflection questions throughout the process (Baker et al., 2012). Learners in high school agricultural education programs have many opportunities to learn through experiences including their instructional labs, SAE projects, and FFA events. Experiential learning will be expanded on in future chapters.

Another application of constructivist theory is **discovery learning**. Discovery learning is simply the process of obtaining knowledge for oneself. It is also a type of problem-solving, is experiential, and uses inquiry learning. Teaching through discovery learning involves presenting questions without readily available answers, problems with no solutions, or challenging situations that encourage learners to create a best-guess answer. Discovery learning works best when learners have some prior experience or background information (Schunk, 2012). More information about discovery learning through labs in high school agriculture programs can be found in chapter 12.

Project-based learning (PBL) is an application of constructivist learning theory. It allows learners to learn by doing, applying ideas, and solving problems. Research on PBL indicates classrooms that use this learning result in better learning outcomes than traditional classrooms. SAEs that consist of research can serve as a type of PBL. PBL involves active construction, situated learning, social interactions, and cognitive tools. To use PBL in the classroom, educators would have learners construct meaning based on their learners' experiences and interactions in the world. Learning can occur in real-world contexts (situated learning). In school-based agricultural education (SBAE) educators, learners, and community members work together for shared understanding (social interactions). Finally, learners make use of cognitive tools which are things like graphs and other visualization and data analysis tools to contribute to their understanding of patterns in data.

Experiential learning, discovery learning, and project-based learning are but three applications of learning theories and can also be considered as pedagogies used in educating learners at all levels. There are other effective pedagogies, such as inquiry-based and problem-based learning, that are developed based on learning theories and are widely used by educators. As you design your instruction for powerful and dynamic learning within your programs, consider how you can make the environment conducive to learning. Keep in mind the individual characteristics of your learners, their motivations and goals, their past experiences, their self-efficacy toward the content, and how you can facilitate not just memorization of content, but deep, transformative learning that translates into more meaningful outcomes in the future. As an educator, you have the power and ability to design your programs to maximize learning for your learners.

Reflective Questions

1. Think about a subject you will be teaching in the near future. What information from this chapter could assist you in structuring the content to create a dynamic learning environment?
2. What do you believe about learning? How will this guide your teaching strategies?
3. Which theories of learning have you seen demonstrated in yourself and in others?

Glossary of Terms

- **learning:** Change in behavior that endures over time and is a result of practice or other experience
- **learning theories:** Theories that explain or predict factors of and concepts that influence learning
- **pedagogies:** Strategies, techniques, practices, and approaches used by educators to facilitate learning
- **4E cognition model:** A model that recognizes learning involves more than just the brain; our bodies, the situation, and interactions within the environment also contribute to learning
- **information processing theory:** A theory that attempts to explain the path of how information is encoded into the memory system
- **social cognitive theory:** A theory of learning that stresses dynamic and reciprocal influences of behavior, environment, and individual cognitive factors where learning occurs by observing others
- **identity exploration framework:** A framework for helping educators factor in individual attributes of learners and their importance to how they learn
- **self-efficacy:** Learner's belief in their abilities or capacity to achieve goals which influences their learning

References

Alvarez, G. A., & Cavanagh, P. (2004). The capacity of visual short-term memory is set both by visual information load and by number of objects. *Psychological Science*, *15*(2), 106–111. https://doi.org/10.1111/j.0963-7214.2004.01502006.x

Baker, M. A., Robinson, J. S., & Kolb, D. A. (2012). Aligning Kolb's experiential learning theory with a comprehensive agricultural education model. *Journal of Agricultural Education*, *53*(4), 1–16. https://doi.org/10.5032/jae.2012.04001

Bandura, A. (1977). *Social learning theory*. General Learning Press.

Bandura, A. (1986). *Social foundations of thought and action*. Prentice Hall.

Casey, B. J., Giedd, J. N., & Thomas, K. M. (2000). Structural and functional brain development and its relationship to cognitive development. *Biological Psychology*, 54, 241–257. https://doi.org/10.1016/s0301-0511(00)00058-2

Casey, B. J., Tottenham, N., Liston, C., & Durston, S. (2005). Imaging the developing brain: What have we learned about cognitive development? *The Trends in Cognitive Sciences*, *9*(3), 104–110. https://doi.org/10.1016/j.tics.2005.01.011

Casey, B. J., Tottenham, N., & Fossella, J. (2002). Clinical, imaging, lesion, and genetic approaches toward a model of cognitive control. *Developmental Psychobiology*, *40*(3), 237–254. https://doi.org/10.1002/dev.10030

Cowan, N. (1995). *Attention and memory: An integrated framework*. Oxford University Press.

Csibra, G. (2007). Action mirroring and action understanding: An alternative account. In P. Haggard, Y. Rossetti, & M. Kawato (Eds.), *Sensorimotor foundations of higher cognition: Attention and performance XXII* (pp. 435–480). Oxford University Press.

Eccles, J. S., & Wigfield, A. (2002). Motivational beliefs, values, and goals. *Annual Review of Psychology*, *53*(1), 109–132. https://doi.org/10.1146/annurev.psych.53.100901.135153

Delazer, M., Ischebefk, A., Domahs, F., Zamarian, L., Koppelstaetter, F., Siedentopf, C. M., & Felber, S. (2005). Learning by strategies and learning by drill: Evidence from an fMRI study. *Neuroimage*, *25*(3), 838–849. https://doi.org/10.1016/j.neuroimage.2004.12.009

Flum, H. & Kaplan, A. (2006) Exploratory Orientation as an Educational Goal, Educational Psychologist, 41:2, 99-110, DOI: 10.1207/s15326985ep4102_3

Gluckman, M., Vlach, H. A., & Sandhofer, C. M. (2014). Spacing simultaneously promotes multiple forms of learning in children's science curriculum. *Applied Cognitive Psychology*, *28*(2), 266–273. https://doi.org/10.1016/j.neuroimage.2004.12.009.

Graham, S., & Weiner, B. (2012). Motivation: Past, present, and future. In K. R. Harris, S. Graham, T. Urdan, C. B. McCormick, G. M. Sinatra, & J. Sweller (Eds.), *APA educational psychology handbook, Vol. 1. Theories, constructs, and critical issues* (pp. 367–397). American Psychological Association. https://doi.org/10.1037/13273-013

Haglund, J., Jeppson, F., & Andersson, J. (2012). Young children's analogical reasoning in science domains. *Science Education*, *96*(4), 725–756. https://doi.org/10.1002/sce.21009

Kaplan, A., Sinai, M., & Flum, H. (2014). Design-based interventions for promoting students' identity exploration within the school curriculum. *Advances in Motivation and Achievement*, *18*, 243–291. https://doi.org/10.1108/S0749-742320140000018007

Krawczyk, D. C. (2018). Reasoning origins: Human development during childhood. In D. C. Krawczyk (Ed.), *Reasoning: The Neuroscience of how we think* (pp. 101–129). Elsevier Inc. https://doi.org/10.1016/C2015-0-05936-6

Lee, J. K., & Kwon, Y. (2011). Why traditional expository teaching-learning approaches may founder? An experimental examination of neural network in biology learning. *Journal of Biological Education*, *45*(2), 83–92. https://doi.org/ 10.1080/00219266.2010.548874

Locke E. A., & Latham G. P. (1990). *A theory of goal setting and task performance*. Prentice Hall.

Locke E. A., & Latham G. P. (2002). Building a practically useful theory of goal setting and task motivation: A 35-year odyssey. *American Psychologist*, 57, 705-717. https://doi.org/10.1037/0003-066X.57.9.705

Maguire, E. A., Frith, C. D., & Morris, R. G. M. (1999). The functional neuroanatomy of comprehension and memory: The importance of prior knowledge. *Brain, 122*, 1839-1850. https://doi.org/10.1093/brain/122.10.1839

Meltzoff, A. N. (1990). Foundations for developing a concept of self: The role of imitation in relating self to other and the value of social mirroring, social modeling, and self practice in infancy. In D. Cicchetti & M. Beeghly (Eds.), *The self in transition: Infancy to childhood* (pp. 139-164). University of Chicago Press.

Muro, M., & Jeffrey, P. (2008). A critical review of the theory and application of social learning in participatory natural resource management processes. *Journal of environmental planning and management*, 51(3), 325-344. https://doi.org/10.1080/09640560801977190

Nie, J., Zhang, Z., Wang, B., Li, H., Xu, J., Wu, S., Zhu, C., Yang, X., Liu, B., Wu, Y., Tan, S., Wen, Z., Zheng, J., Shu, S., & Ma, L. (2019). Different memory patterns of digits: a functional MRI study. *Journal of Biomedical Science*, *26*(22). https://doi.org/10.1186/s12929-019-0516-y

Piaget, J. (1936). *Origins of intelligence in the child*. Routledge & Kegan Paul.

Piaget, J. (1970). *Genetic epistemology*. Columbia University Press.

Pintrich, P. R., Cross, D. R., Kozma, R. B., & McKeachie, W. J. (1986). Instructional psychology. *Annual Review of Psychology, 37*, 611-651. https://doi.org/10.1146/annurev.ps.37.020186.003143

Pouw, W.T.J.L., van Gog, T. & Paas, F. (2014). An Embedded and Embodied Cognition Review of Instructional Manipulatives. *Educational Psychology Review, 26*(1), 51-72. https://doi.org/10.1007/s10648-014-9255-5

Rizzolatti, G. (2005). The mirror neuron system and its function in humans. *Anatomy and Embryology, 210*, 419-421. DOI 10.1007/s00429-005-0039-z

Schunk, D. H. (2012). *Learning theories: An educational perspective* (6th ed.). Pearson.

Shuell, T. J. (1986). Cognitive conception of learning. *Review of Educational Research*, 56(4), 411-436. https://doi.org/10.3102/00346543056004411

Sinai, M., Kaplan, A., & Flum, H. (2012). Promoting identity exploration within the school curriculum: A design-based study in a junior high literature lesson in Israel. *Contemporary Educational Psychology, 37*(3), 195-205. https://doi.org/10.1016/j.cedpsych.2012.01.006

Skinner, B. F. (1976). *About Behaviorism*. Vintage Books.

Steier, R., Shapiro, B., Christidou, D., Pierroux, P., Davidsen, J., & Hall, R. (2019). Tools and Methods for '4E Analysis': New lenses for analyzing interaction in CSCL.

Sweller, J. (1994). Cognitive load theory, learning difficulty, and instructional design. *Learning and Instruction*, *4*(4), 295–312. https://doi.org/10.1016/0959-4752(94)90003-5

Sweller, J., van Merriënboer, J. J. G., & Paas, F. (1998). Cognitive architecture and instructional design. *Educational Psychology Review*, *10*(3), 251–296. https://doi.org/10.1023/A:1022193728205

Vygotsky, L. S. (1978). *Mind in society: The development of higher psychological processes*. Harvard University Press.

Waterman, A. S. (1989). Curricula interventions for identity change: Substantive and ethical considerations. *Journal of Adolescence*, *12*(4), 389–400. https://doi.org/10.1016/0140-1971(89)90062-6

Winne, P. H. (1985). Cognitive processing in the classroom. In T. Husen & T. N. Postlethwaite (Eds.), *The International Encyclopedia of Education* (Vol. 2, pp. 795–808). Pergamon.

Wirebring, L. K., Lithner, J., Jonsson, B., Liljekvist, Y., Norqvist, M., & Nyberg, L. (2015). Learning mathematics without a suggested solution method: Durable effects on performance and brain activity. *Trends in Neuroscience and Education*, *4*(1), 6–14. https://doi.org/10.1016/j.tine.2015.03.002

Wittrock, 1986: Please replace Wittrock, 1986 to Brophy, J., & Good, T. L. (1986). Teacher behavior and student achievement. In M. C. Wittrock (Ed.), Handbook of research on teaching (3rd ed., pp. 328-375). New York: Macmillan.

Wood, R., & Bandura, A. (1989). Social cognitive theory of organizational management. *Academy of Management Review*, *14*(3), 361–384. https://doi.org/10.2307/258173

Woollard, J. (2010). *Psychology for the classroom: Behaviourism*. Routledge.

3. PRINCIPLES OF TEACHING AND LEARNING

Kasee L. Smith

Setting the Stage

Carmen set out to evaluate three teachers in her school. As a new vice principal, her job was to examine the quality of teaching for educators in her building. She walked into three classrooms that morning and observed the following:

- Mr. Patterson had his lesson plans completed and organized by date, time, and course. He worked to ensure that everyone he instructed was presented with opportunities to learn. He had an interesting lesson opening, and made sure learners knew the objectives for the day before they begin. Learners were guided through carefully planned lessons which included opportunities to demonstrate knowledge, relearn information, and work toward stated goals.
- Ms. Diaz loves her content area and is consistently enthusiastic about the things she expects others to learn. She worked to make sure she clearly described concepts and occasionally used humor to lighten the mood and create a positive learning environment. She integrated new learning activities into her instruction and provided opportunities for learners to receive positive reinforcement for completing tasks and learning concepts. While enthusiastic, she made sure her learners remained on task and worked through the material they should be completing.
- Ms. Novak centered learners in her approach to instruction. She focused on motivating learners to want to learn. She developed and delivered engaging lessons that allowed learners to connect with the content on a deeper level. Her focus was on creating instruction which helped learners retain information and apply content to other situations.

Objectives

By the end of this chapter, learners will be able to:

- Define teaching and learning.
- Explain the purpose for following principles of teaching and learning.
- Describe principles for effectively organizing content.
- Describe principles for effectively delivering content.
- Describe principles related to learning.
- Share principles of teaching and learning in practical scenarios.

Introduction

Learners naturally make appraisals of the quality of educators. Most people can list the characteristics we use when describing good and bad educators, which invites the question: what makes someone a good educator? Which of the teachers Carmen observed would you choose to learn from? Do you think these educators are *good* at their job? What characteristics do they exhibit that indicate their potential for success in the classroom?

Now, think for a moment about the best educator you have ever had ... it might be from elementary school, high school, a college professor, or someone who taught you a specific hobby or life skill in an informal setting. Next, think about what they did that made them a good educator from your perspective. Were they passionate about the content? Did they encourage you personally or academically? Did they care about you as an individual? Make a mental list of their teaching characteristics or jot them down on a piece of paper.

Is the best educator you've ever had similar or different from the three educators we examined at the opening of the chapter? What commonalities exist between the three educators in the scenarios and your ideal educator? The simple truth is, there are many ways to be a good educator. Every learner is different, and each learner has different learning needs. Every educator is different too. One of the challenges to becoming an educator is knowing that there are people who do the job well. You might be thinking, "I don't think I could be like the educators in my scenario, and I don't think I'll ever be as good as the best educator I've ever had." That's okay. Our role as developing educators is to find our own path and to develop a set of teaching skills that are uniquely matched to us.

The principles of teaching and learning are the foundation upon which great educators are built. These concepts relate to organizing content for learners, delivering instruction, and understanding how learners learn. In our original scenarios Mr. Patterson focused on the principles of teaching related to organizing content, Ms. Diaz focused on the principles of teaching related to delivering instruction, and Ms. Novak focused on the principles of student learning. In this chapter, we will examine the principles of teaching and learning and discuss where they came from, how they function in a learning environment, and why they are important.

Overview of the Principles of Teaching and Learning

Educational researchers have been working to define the principles of teaching and learning for decades. In fact, many of the earliest philosophical conversations in most cultures were related to determining how information should be taught and how people should learn. So, before we can discuss the principles of teaching and learning, let's pause to define both teaching and learning. Teaching is the process of mutual exchange and/or relationship among an educator, learners, and curriculum or society (Dewey, 1938). Learning is the process of acquiring and transforming information into knowledge which can be assimilated, transformed, and processed by the learner (Sousa, 2010).

Darling-Hammond and Bransford (2007) suggested that to have a working ability to teach, a novice educator must be able to understand learners, curriculum content, and how to teach. While there are many ways to think about teaching and learning, the core principles of teaching and learning are the same. For the purpose of this chapter, we will examine the principles of teaching and learning in three contexts:

- **Principles of Organizing Content.** These principles deal with how educators transform information in the world into curriculum that is ready to be shared with learners.
- **Principles of Delivering Instruction.** These principles provide insight for educators related to how they interact with learners in the learning environment.
- **Principles of Student Learning.** These principles are related to learners perceiving instructional cues, receiving information, and assimilating new learning into their current knowledge base.

Organizing Content

Have you ever encountered educators who made you wonder if they were organized and ready to teach? Did you ever worry that an educator was ill-prepared to deliver content? The first step in preparing to educate learners is to organize the content you will teach. We will discuss processes for organizing content in detail in chapter 7, "Planning for Effective Instruction." From a broader perspective, there are important principles of teaching that can help guide thoughts around preparing content for learners.

Principle 1: Subject matter should be organized in a sequence and logically from simple to more complex.

Organizing content is best done when we meet learners at their current level of knowledge, then increase the complexity of concepts, while making sure to tie new information to what learners already know. Organizing the content means first organizing the subject matter by listing the items learners are expected to know at the end of the course. These learning outcomes, often called objectives, should be sequenced to present learners with more foundational concepts first. This can allow learner knowledge to build based on the information they have already acquired in the class. This concept is typically referred to as scaffolding, which makes sense if you think for a moment about the way a scaffold works. To reach higher levels on a scaffold, we need to climb up each layer below. The same concepts are true with learning. Learners should learn more basic concepts first, then they can increase their level of cognition and build upon their new knowledge. A large-scale example of the concept of scaffolding is learning to read. Before someone can read, they first need to be able to recognize each letter, once letters are recognizable, the sound each letter makes can be added to the knowledge, once sounds are understood for each letter, small strings of letters can be added to form short words. Gradually, as understanding of letter combinations and phonics rules increase, someone can scaffold their knowledge to read complicated words.

Knowing student knowledge level and prior experiences is critical to successfully integrating principle 1. How can we discover what learners already know? A good first step is to ask those who know the learning environment and background of the learners what the knowledge strengths are for the group of learners. Becoming an excellent educator means understanding our learners, their interests, and the prior experiences they might have. With this knowledge of learners comes a clarity in the level of knowledge they already possess. A possible strategy for gathering learners' knowledge is a preassessment of knowledge and experiences. Conducting a preassessment can provide the needed information for the educator to know where the scaffold should begin. Perceptive educators also gain skills to interpret learner understanding and level of knowledge based on the questions learners ask and the responses they give during instruction.

How do we sequence instruction once we have background information on learner content knowledge? An important question to ask is are there sections of the content which will require learners to know knowledge from another section of the content? Some instruction includes knowledge that learners need to build on to grasp more complicated concepts. The foundational concepts should therefore be instructed before the more complicated components. The same is true for objectives for a lesson plan or a series of learning sessions.

While learning is the most important consideration for sequencing instruction, there may be other items to consider in planning a series of instructional lessons. Thinking about how the content aligns with seasons or other local, state, or national events can provide more opportunities for learners. For example, it makes sense to align instruction on transplanting in a greenhouse with the actual greenhouse transplanting schedule or to plan instruction in public speaking to align with a local or state competition. Planning long-term instruction is something that can give learners a foundation upon which to build their content knowledge.

The sequence of instruction should not be a secret to learners. Sharing the sequence with learners allows them to create a visual roadmap of where the instruction will go next and allows them the opportunity to look back on the knowledge they have gained. Giving learners access to the sequence for both the entire series of instruction and within smaller instructional units can provide the clarity learners need to stay focused on the objectives of the instruction.

What does Principle 1 look like in action?

- In a greenhouse management session: Delivering a unit on parts of the plant before a unit on asexual propagation where parts of the plant will be referenced.
- In an animal science session: Ensuring learners know what the types of injections are before a lab on giving injections.
- In a leadership workshop: Discussing the components of a speech prior to having learners compose a speech of their own.

Once we know the flow and sequence of instruction, we can begin to organize individual learning sessions. Each learning session should be organized in a way that allows learners to fully engage with the content and demonstrate their understanding. Principles 2 through 8 are involved with each step of instruction. These principles are built upon the nine events of instruction as developed by Robert Gagne in 1965. Those nine events are:

- Gain attention
- Stimulate recall of prior knowledge
- Inform learners of objectives
- Present stimulus
- Provide learner guidance
- Elicit performance
- Provide feedback
- Assess performance
- Enhance retention and transfer

Let's examine the principles behind each of these events of instruction and how organizing content can be accomplished by paying attention to the intent of each of these steps.

Principle 2: Learners will not learn unless they find interest in the subject matter; they learn more when they understand why the content is important to their life.

Imagine this scene: You are sitting in the back of a traditional high school classroom when learners come into class. Several are talking to each other, a few more are paying attention to their phones, and a couple even come into the room and put their heads down on the desk. The bell rings and the teacher says, "Start by reading the handout on genetics, then we will discuss." The educator has obviously planned for the day, but are learners engaged and ready to learn? Without learners shifting their focus to the learning session and topic, very little the educator does will stimulate learning. Learners must be attentive to the educator and to the content for learning to occur.

How can we gain learners' attention? There are many ways to ensure learners pay attention to the content. Activities designed to stimulate learner attention at the beginning of a learning session are commonly called interest approaches, though they can also be called anticipatory sets, warm-ups, start tasks, or bell ringers. These attention-stimulating activities are designed to help learners engage with the content and the educator. Interest approaches can be designed to simply gain student attention but can also extend into allowing learners to connect their prior experience to the current content or compel them to want to learn more about the topic. Interest must be gained at the beginning of each learning session and at the transition between units of instruction. Gaining content-specific attention is key for learners to be able to engage with the content and to be ready to learn.

You might recall being in a class and thinking, "When will I ever use this?" Being able to answer this question for learners is a core component of instruction. Learners pay more attention to things that matter to their lives currently and in the future. Master educators look for opportunities to tie the content to the lives of their learners through examples or explanations that are relevant. The goal for educators is to create a need. This switch for a learner is the difference between learning content to complete a session or get a score and learning content because it matters in the real world. Our content provides many opportunities for learners to reach this *felt need* to learn real-world content. The educator's role is to explain why each subject in the curriculum is important to learners' lives.

What does Principle 2 look like in action?

- In a water quality workshop: Sharing stories of waterborne illnesses at the opening of a session designed to help learners know how to properly test drinking water.
- In a small engines session: Showing learners an engine that is running poorly and asking learners what they think might be wrong at the beginning of a lesson on troubleshooting.
- In a biology session: Showing learners highly magnified images of animal cells dividing before a lesson on mitosis, then telling learners that understanding this process can help them better understand how both regular growth and cancer works.

Principle 3: Learners learn best when they can tie new knowledge to prior knowledge and experiences.

Can you recall the last time a song made you think of an experience? One of the most remarkable things about humans is our ability to tie new knowledge to things we already know, like a song. The process of learning new information requires that we incorporate new information into that which we already know, a process called assimilation. For assimilation to be most effective, the prior knowledge or experience should be brought back to mind prior to being exposed to new information.

Educators can help learners assimilate new knowledge by asking learners questions to recall information they already know. This questioning process may be as simple as asking learners to recall what was done in class the prior day or as complicated as asking learners to recall an adjacent concept then drawing an analogy to the new content to be learned. When learners recall prior knowledge, they make their brain more ready to receive new information. Recalling prior knowledge allows learners to anchor new knowledge to concepts they already know and can recall. It is similar to pulling a file folder out of a filing cabinet before putting a new paper in it; when a new paper is added to the folder, it is stored with all the other related concepts. In most instances, this method of learning is more effective than learning through repetition or rote memorization because it helps develop neural pathways that can be retained. The more times learners recall information, the more likely they are to be able to retain the information for long periods of time (Rock, 1957).

What does Principle 3 look like in action?

- In a floral design workshop: Posting a slide as learners arrive for a second class with the prompt, "Please list all of the color schemes you remember from yesterday's lesson."
- In a metal fabrication shop: Asking learners to share times when they have seen or built something that was not square. Holding a group discussion about the implications of a project not being square.
- In a leadership session: Asking learners to recall times when conflict occurred on a team and the outcome of the conflict prior to discussing methods for conflict resolution.

Principle 4: Learners best gain knowledge when they know what they are expected to learn.

Do you know the most asked question in all K-12 classes in the United States? Believe it or not, the most commonly asked question is, "What are we going to do today?" Why is this the most asked question? Learners are curious and wanting to know what they will learn comes naturally to them. Being able to satisfy curiosity is a good reason to let learners know what they will learn, but sharing learning outcomes with learners is so much more than that.

Of course, for learners to know what they are expected to learn, the educator needs to know what they expect learners to learn. Understanding the objectives of your instruction is a critical component to organizing content and must be the starting point when planning long-term or short-term instruction (Darling-Hammond & Bransford, 2007). Most often, the expectations for learning are called the objectives when the educator is planning and become less formal when presented to learners as a learning outcome. Whether formal or informal, learning expectations should be stated explicitly in measurable terms. An example of an objective and learning outcome include:

- Objective: Learners will be able to explain the importance of personal protective equipment (PPE) when working in a laboratory environment.
- Learning Outcome: You will explain why we wear PPE in the lab.

We might wonder, "If the educator knows what the learners will be learning, isn't a simple agenda of what they will do enough?" The answer from educational researchers is a resounding no. When a learner knows what they are expected to learn, it has implications for the way content is organized and delivered. One of the biggest benefits to sharing the learning objectives with learners is that they can store new information along with information they already know (Gagne, 1965). To go back our example, when the educator shares with learners they will be learning about PPE, their minds automatically recall their previous experiences and knowledge about PPE. This recall process prepares the brain to create neural networks to the new information. Those networks are critical in learners being able to recall new information later (Sousa, 2010).

As educators, wouldn't it be nice to have a room full of learners pushing toward learning what we expected them to learn? Another reason for sharing objectives and learning outcomes with learners is that it gives learners more control and responsibility for their own learning (Darling-Hammond & Bransford, 2007). When learners know what they should be able to do, they are more likely to remain on task, ask more questions, and help guide the pace of instruction (Gagne, 1965). Sharing objectives with learners also allows them to self-evaluate their knowledge. Master educators share objectives at the beginning of each instructional session, then they revisit the objectives at the end of the session to assess learner progress. Gathering evidence related to the objectives allows educators to determine if content is mastered, if there are misunderstandings, or if learners need more guidance to grasp the concepts.

We might wonder, what do educators do for classes where learners are working on project-based learning assignments? Even in laboratory settings or with self-paced work, learners can benefit from knowing what they are expected to learn for the day. In a highly individualized setting, this might mean letting learners set their own learning objectives, and assessing their performance based on their own goals. Regardless of the type of learning environment, sharing learning outcomes with learners is perhaps the principle that is easiest to implement and that yields the biggest result for learning. The concept is so important that many evaluation tools include a rating for educators regarding their ability to share learning outcomes with learners.

What does Principle 4 look like in action?

- In a wildlife management workshop: Asking learners to copy down what they should be able to do by the end of class, then at the end of class, asking learners to check off the objectives they feel like they accomplished.
- In a mechanics laboratory: Having each learner write what they plan to learn during lab time on a sticky note and placing on the door on the way to the lab. At the end of class, have learners collect their sticky note and write how they accomplished their objective on the sticky note to turn in for work points

Principle 5: The method best suited to the content must be used to maximize learning.

There are many different instructional methods educators use to stimulate learning. From lecture to field trips to inquiry-based learning methods, educators must help learners gather knowledge. This process is referred to by Gagne (1965) as "presenting the stimulus." Sometimes, instructional methods are referred to as "teacher-centered" or "learner-centered" and there are many different opinions for which method to use (Norman & Spohrer, 1996). Master educators have a wide variety of knowledge of instructional methods, and they present information to learners in the format that is best for learners.

There is more than one correct way to present the learning stimulus. How do educators choose an instructional strategy to match the content? The first question to ask is "What are the objectives?" Each objective should be written in a way that helps the educator envision how learners will be assessed. For example, if the objective is for learners to describe the influence of light intensity on plants, it might be appropriate for learners to inquire into the concept by designing an experiment to test light intensity. If, by contrast, the objective is to list the types of greenhouse lighting, an experiment is not logical, as the outcome is simply to list. In this case, another instructional strategy (like lecture, or collaborative learning) might be more appropriate.

The next question to ask when determining how to present the stimulus to learners is "How can we maximize learner engagement?" Many can recall being bored in a class. Most likely, boredom stems from not being given enough opportunity to engage during the instruction. Learners better grasp information when they have the opportunity to play an active role in the learning, which will be discussed in more depth in principle 15. Other considerations for presenting the learning stimulus to learners are how learners in the class prefer to interact with the subject matter and each other, how much time is allotted for learning the content, and the comfort level of the educator in using varying instructional strategies.

What does Principle 5 look like in action?

- In a floral design workshop: Giving each student a different description of a color scheme and asking them to find an example from a magazine to share with the class. Once they have found their examples, learners can share their findings with others who collect the information about the color schemes in their notes.
- In a plant science session: Having learners set up an inquiry-based learning experiment to determine the influence of different light intensities on plants. Learners keep detailed observations and, at the end of the experiment, share their findings and collaborate with others to develop an understanding of light intensity.

Principle 6: Learners must be given opportunities to practice and to receive feedback to clarify their understanding.

Have you ever thought you understood something, only to find out later your knowledge was missing a key component? Imagine this scenario: A educator demonstrates a fillet weld to a group of learners. They then give each learner two pieces of metal and tell them to go complete the weld. Learners are scored on their first attempt. What is wrong in this scenario? If you wonder why the educator in this scenario did not allow learners the opportunity to practice and improve, you are well on your way to understanding our next principle.

The next principle of teaching related to organizing content is the process of allowing learners to identify and fill in the gaps in their understanding prior to the final (summative) assessment of their knowledge or skills. Learners should be prompted to demonstrate their knowledge under the guidance of the educator, who can provide additional information or correction while the student is still forming their full understanding of the content. This process is repeated as a loop which includes providing learning guidance, eliciting performance from the student, and then providing feedback. The process is also referred to as formative assessment because it happens when the knowledge is still forming in the student's mind (Harlen & James, 1997).

One of the most important parts of formative assessment is allowing learners to self-assess performance. Learners should be asked questions to recall their knowledge in order for the feedback to be most effective. If a student brings an educator a project and asks if the project is good, master educators will most often ask the learner questions about the project in a process which can lead to the learner assessing the work on their own. For example, rather than telling a learner their travel speed was incorrect on a weld, a master educator might point out a section and ask the student, "What do you think about the travel speed here?"

Completing the feedback process helps educators make decisions about when to move on to a new concept or where common misunderstandings might still exist. By asking learners to demonstrate their understanding before the final assessment, learners are given more opportunities to recall information, which can lead to both a deeper understanding of the subject and higher retention for the concepts in the future.

What does Principle 6 look like in action?

- In a beginning welding session: Allowing learners to practice a weld and receive feedback from the instructor between attempts before completing a final weld for a grade.
- In a session on teamwork: Allowing learners to struggle to find common ground on a topic, then asking them to discuss as a group why common ground was difficult to find.

- In an agribusiness workshop: After teaching learners about interest rates, giving learners a worksheet that includes questions on calculating interest rate. After learners complete the worksheet, the educator notices that many learners failed to correctly convert percentages. The educator hands back worksheets with comments and reteaches how to convert percentages to the class the next day.

Principle 7: Educators must assess student knowledge to acquire evidence of understanding.

An important component of organizing content is to know what learners are expected to know at the end of the learning session. Another important component of organizing content is to know how you will gather evidence that learners can meet those expectations. Without assessment, there is a lack of formal evidence that learning took place.

Providing feedback as discussed in principle 6 allows learners to complete formative assessment, but learners should also be asked to demonstrate their knowledge after all the learning within an instructional unit is complete. This demonstration of knowledge is called a summative assessment. A summative assessment is a record of learning that allows educators to make decisions about moving on with the sequence of instruction or continuing to provide guidance to students. Because learning is sequenced to build on itself, completing an assessment is a way of ensuring learners are ready for the next content area.

Summative assessments can take many forms depending on the objectives. For performance-based learning objectives, learners may be assessed using a rubric designed to allow the educator to assign a score based on performance. Knowledge-based objectives might be assessed using a traditional exam, or through a project, paper, or other activity. In agricultural education, both knowledge and performance skills can be assessed through integrating Leadership Development Events (LDEs) or Career Development Events (CDEs) into the learning as an authentic assessment of knowledge and/or skills. For example, if your food science class includes objectives related to developing a product label, your assessment could mirror the National FFA Food Science CDE Rubric and learners could attend a district, area, or even state event as their summative assessment.

What does Principle 7 look like in action?

- In an introduction to agricultural mechanics session: Giving learners an exam which allows them to identify hand tools after learning about hand tools.
- In a food science class: Using the National FFA Food Science CDE Rubric to assess student ability to develop a food nutrition label.

Principle 8: For learners to retain information, they must be repeatedly prompted to recall concepts.

If a student cannot remember what they learned, the learning process is not complete. Retention is the most difficult part of learning, as it requires information to be stored in a way that allows recall (Rock, 1957). Some of the principles we have already discussed (recall of prior knowledge and informing learners of the objectives) are important to storing information for retention (Rock, 1957), but there are other considerations to think through as well.

For a learner to be able to retain information, the information must be recalled. Each time a learner is asked to recall information, they are more likely to retain the information for a longer period of time (Tinto, 1990). Educators can help this process by looking for opportunities to recall information from prior lessons, units or even courses. Educators can also help this process by allowing learners to recall information in other content areas and relate it to their agricultural education classes.

What does Principle 8 look like in action?

- In a meat science workshop: Asking learners to list the bones inside meat cuts and which other bones they connect to in a complete carcass, then asking them to identify the bones on the live animal.
- In a greenhouse session: While pinching plants in the greenhouse, quizzing learners on the identifications of the plants growing to help with plant ID recall.

Delivering Instruction

Being a good educator means more than just having lesson plans ready to go, it means being prepared to deliver instruction in a way that allows learners to connect. While well-organized content is the foundation of any class session, if the educator cannot make a connection with learners, even the best-prepared lesson might not result in student learning. There are several principles of teaching that are not related to how content is delivered but are more closely related to how the educator presents themselves and the content to learners.

What are the characteristics that make an educator good at delivering instruction? If you think back to your favorite educator, you will likely be able to identify some of the same characteristics educational researchers have identified. One of the most widely recognized studies on educator characteristics and the relationship to student achievement is the seminal 1971 study of the relationship between teacher behaviors and student achievement conducted by Rosenshine and Furst. These researchers conducted a meta-analysis of numerous studies which examined characteristics of classroom teachers and organized the behaviors into a set of best practices for educators. There were eleven educator characteristics which were found to correlate with student achievement, including:

- Clarity
- Variability
- Enthusiasm
- Task-oriented and/or businesslike behaviors
- Criticism
- Student opportunity to learn criterion material
- Use of student ideas and general indirectness
- Use of structuring comments
- Types of questions
- Probing
- Level of difficulty of instruction

The next set of principles of teaching are directly related to the five most prevalent educator characteristics identified by Rosenshine and Furst (1971) which are: clarity, variability, enthusiasm, task-oriented and/or businesslike behaviors, and student opportunity to learn criterion material.

Principle 9: Learners learn only when the educator provides clarity in their instructions, their content, and their expectations.

Rosenshine and Furst (1971) found that the single most prevalent classroom teacher characteristic related to student achievement was the ability of a teacher to teach with clarity. Clarity is the concept that learners can comprehend what the educator intends them to know, do, or think. It is not enough for educators to understand the content, we must also be able to translate our own knowledge into a language learners can grasp. We must then assemble instructional strategies, practice giving directions to activities, and choose assessments that are also clear to learners.

Educators that strive for clarity spend less time answering questions from learners related to interpreting what the educator wants them to do. Educators who demonstrate clarity are more likely to use a linear path for describing content and assignments, and they understand the content well enough to answer questions in a straightforward manner. Educators who possess clarity are also more likely to be organized and prepared for class.

There are several ways educators can improve clarity. First is to prepare a detailed plan for each lesson. Writing a detailed plan provides time to think through not only the content, but how the educator will move through the flow of the lesson. Including details in the plan such as how learners will move into groups or where assignments will be turned in are indicators that the logistics of a lesson are well-conceived and will be delivered with clarity. Other options for improving clarity as an educator include practicing the lesson and receiving feedback from a mentor before delivering the lesson; establishing and following classroom procedures for housekeeping tasks like turning in papers or moving into groups; and providing instructions at the top of each handout or worksheet.

What does Principle 9 look like in action?

- In a mechanics workshop: Educator highlights places where learners often make mistakes when describing a step-by-step plan to change the wire spool in a MIG welder. Learners have a handout which they can follow along and read the steps in order to the educator during the demonstration. When asked a question, the educator clearly and correctly explains the answer.
- In a learning environment: Educator refers to their plan when breaking learners into groups for a review. The instructions for moving into groups were planned out in the lesson plan and the educator clearly explains the rules to learners before the review.

Principle 10: Each student learns differently; good instruction has enough variability to maximize learning opportunities for all learners.

Variability, at its core, addresses inclusivity in a learning environment. Varying our delivery and our assessment as educators helps ensure every student is included in every learning session. All learners learn differently. Varying our behaviors as educators provides the opportunity to align some aspect of the learning environment to the learners' preferences for bringing in or transforming information. In addition, by varying instruction, educators can prevent learners from becoming bored and losing interest or motivation toward the content. Variability means using different instructional strategies, types of assessment, instructional materials, and active learning strategies in every learning environment. As mentioned in principle 5, there are many ways to present information to learners. Having variability means using many of these strategies with the same group of learners.

Educators who strive for variability typically have more engaged learners and can reach learners with diverse learning needs. The more variation you can integrate, the more likely you will be to instruct classes that align with student learning preferences, giving every student a chance to succeed. Some options to increase variability as an educator include implementing a new instructional strategy, type of formative assessment, instructional technology, or type of questions on tests; varying the learning environment by hanging different instructional aides in the classroom or occasionally changing the flow of your class sessions; and bringing in a guest speaker to vary the person delivering the information to learners.

What does Principle 10 look like in action?

- In an animal science workshop: Asking learners to create clay models of digestive systems, the next day, allowing learners to complete a dissection of a ruminant digestive tract, and the following day, allowing them to research digestive disorders on the internet to share with the class.
- In a sustainability workshop: Putting students in groups for discussion, then having them create Venn diagrams of the discussion, then allowing them to hear from a specialist in the area of sustainable production.
- In a biology session: After playing one online game for reviewing content all year, switching platforms and asking learners to use a new game for the review.

Principle 11: Enthusiasm toward a content area is contagious. When the educator is excited to teach content, learners are more likely to be excited about learning the content.

Have you ever watched someone teach something they were really excited about? Enthusiasm is contagious. If educators are passionate and excited about what they are teaching, researchers have found that learners will increase their engagement, pay closer attention to the topic, and achieve more than when educators display less passionate about a subject (Keller et al., 2013).

How will our learners know we are excited about a topic? Movement, gestures, and changes in voice can signal enthusiasm, as can the number of personal stories and facts shared on a topic. By raising our energy level learners notice our enthusiasm about the topic. You might be thinking, "But what if I'm NOT enthusiastic about the subject matter?" Consider utilizing a guest speaker who is passionate about the topic or finding a video, story, or news article that paints the topic in an engaging or exciting light.

What does Principle 11 look like in action?

- In a horse management workshop: Educator shows pictures of hoof deformities, goes into depth on the types of corrective shoeing, and shares numerous examples of the importance of hoof care as a component of equine health management.
- In a landscape design session: Educator develops numerous ways for learners to remember each plant, has learners collect plant samples, and shares stories of what each plant is used for in a landscape.

Principle 12: Learners learn best when educators are businesslike and approach teaching as a profession.

Educators who are task-oriented or businesslike remember that the goal of teaching is for learners to understand content. Educators with this characteristic follow a code for professionalism which includes maintaining high expectations, setting clear behavioral guidelines for the learning environment, and challenging learners to work to their full potential. By setting an expectation that learners will behave and challenge themselves, most learners will work to meet those expectations. Being businesslike does not necessarily mean that educators should not allow fun or enjoyment in the learning environment, it simply means that student learning is the task at hand. Educators who are informal and playful can still meet a businesslike principle if the main focus of their work is helping learners grasp concepts.

Businesslike or task-oriented educators work to ensure that transition time (time between learning activities) is well managed. Task-oriented teachers also follow a schedule for the instructional session which allows them to meet the objectives in the time allotted. These educators often model laboratory points after real-life work situations and look for opportunities to incorporate critical thinking and application of knowledge.

What does Principle 12 look like in action?

- In a fabrication session: Educator provides positive reinforcement to learners who complete their cleanup work by following the cleanup chart to meet the expected requirements.
- In a food science workshop: Educator ensures students are learning about the molecular structure of foods. While eating the food they make, learners are assessed on their knowledge gain before returning to the food lab.

Principle 13: Learners can learn only when they are given opportunities to practice knowledge and receive immediate assessment.

The amount of time a learners spend practicing and reinforcing knowledge has a direct influence on their level of achievement. Practice opportunities, followed by immediate feedback, directly aligned to objectives are critical. Educators need to provide learners with as many opportunities to learn material related to the objectives as possible. To maximize the opportunity for learners to learn, educators can structure their class to minimize unneeded information and off-task time. The first thing educators can do to provide opportunities to learners is to create a structured learning environment. When a learning environment has procedures for completing logistic tasks (e.g., how to begin a session, roll call, picking up and turning in papers, using a hall pass, getting supplies), the time available for learners to practice is increased.

Educators can also increase student opportunities by providing supplemental or required learning that happens outside of the learning session. This might include sending reading material or application practice home with learners to complete as homework. In a laboratory-based class, this may include extending open lab hours when learners can work on projects outside of the structured session.

What does Principle 13 look like in action?

- In a natural resources class: Learners will be assessed on their ability to determine water pH. The educator gives each student a handout with instructions, test strips, and a practice worksheet that allows them to test water at five locations of their choice, then tests the samples to see how close the student was to the actual result. Learners are given scores before the test.
- In an agricultural structures class: Learners will be assessed on their ability to measure and cut pieces for a project. The educator has learners cut boards at different lengths as a practice and measures each board to see how close the student was to the correct length. Learners are individually coached about how to adjust measurements if needed.

Student Learning

Knowing what it takes to organize and deliver instruction is just one part of being able to understand how learning occurs. Each of the previous thirteen principles focus on the role of educators in the learning process, but there is another side to the learning equation: the learner. Learners are unique individuals, what works to teach one may not work for another. To help you think about how learners learn, there are a few more principles of learning to share.

Principle 14: Every learner is unique and has unique learning needs.

No two learners are the same, and no two educators are the same. That means for every student and every educator combination, the relationship is different. Most classroom teachers and many educators in nonformal learning environments will encounter learners who have Individualized Education Plans (IEPs) or other accommodations under a Section 504 designation. In these situations, educators are legally bound to be flexible and allow learners an opportunity to succeed. Even outside IEP and 504 situations, learners learn best when educators acknowledge and accommodate students' unique learning preferences and develop an individualized plan for learning. It may seem farfetched to think that each student should have a slightly different experience in the learning environment, but it is the mark of a master educator to treat each learner according to their needs. The process of varying instruction for individual learners is called differentiation.

How can we accommodate every learner? First, masterful educators make the effort to know their learners as people. By understanding what learners like and dislike and what they believe their talents and shortcomings are, educators are better able to motivate learners. From an instructional standpoint, educators create differences in content, process of learning, or assessments for learners based on their readiness to learn, interests, and individual learning preferences. Some learners move more quickly through the content because of prior knowledge or increased cognitive ability while some may require more intervention. How an educator plan for unique differences matters for learner achievement.

When educators are working toward differentiation, instruction is flexible. Educators make decisions for what learners need based on an ongoing assessment of progress and careful attention to changes in learner knowledge and behavior. Variability, as mentioned in principle 10, is critical to differentiating instruction.

What does Principle 14 look like in action?

- In a veterinary medicine class: The educator has one student with a 504 requiring modifications on time and level of content and another student who is very gifted enjoys video creation. The educator modifies most assignments for the student with the 504 to allow the student to do only lower cognitive level tasks. The gifted student often finishes assignments and project early, and the educator has commissioned the student to create videos of the Veterinary Science CDE practicums to use in their practical lab procedure unit.

Principle 15: Learners must be motivated to learn.

If you ask most high school students what they would like to do with their time, very few would respond that they would prefer to be in school five days a week. Although school and learning are critical for brain development and social development, many learners would prefer not to spend their time in a formalized learning environment. Educators serve as the motivators for learning in both formal and informal settings. Motivated learners take the initiative to learn content, complete assignments, and assess their own progress toward learning objectives. Motivation is also an important predictor of knowledge and overall achievement. Learners who are more motivated to learn persist through challenges, put in more effort, are more likely to engage with the content, and perform better in classes and on standardized tests.

There are two main classifications of student motivation, intrinsic and extrinsic (Rotter, 1966). Intrinsic motivation comes from the satisfaction of the activity itself rather than an outside reward. This type of motivation is more likely to persist across content areas and directly correlates with increased effort. Extrinsic motivation is driven by an external reward like grades. Both types of motivation play a role in learning.

How can educators motivate learners? There are several things successful educators use to motivate learners:

- Give positive reinforcement of learning
- Set goals for learning and celebrate accomplishment
- Be excited about student performance
- Give learners choices about the content/topic
- Variability
- Relate topics to student life
- Help learners see their progress
- External rewards (e.g., praise, progress charts, tangible items) (Rotter, 1966)

What does Principle 15 look like in action?

- In a youth STEM workshop: Learners are given the opportunity to choose their topic for a science fair project. The educator knows a student is passionate about plants and suggests the student use the school greenhouse to complete the project, then informs the class that the top scoring project in each category will advance to a district-level event.

Principle 16: Learners learn more rapidly when they receive detailed feedback for their efforts immediately after demonstrating their knowledge.

How do we feel when someone tells us did the right thing? Positive reinforcement is one of the best ways to motivate people (Walker & Buckley, 1968). Providing positive feedback to learners can help motivate additional learning and stimulate desire to continue learning the subject matter. Providing developmental feedback can help learners correct errors in their knowledge or skill before those misunderstandings are firmly rooted in their mind.

What are the best tips for providing feedback to learners? To begin, researchers recommend being as specific as possible about the actions that led to the feedback. Giving learners the comment "Great job!" is not nearly as effective as saying "You did a wonderful job creating balance and form in this floral design." The more specific feedback is to learners, the more likely it is that they will repeat the positive behavior. Keeping developmental feedback focused on the learning objectives ensures that we are avoiding criticism that could be taken personally. The next tip is to give feedback as quickly as possible, which has a direct influence on increasing the effectiveness of the feedback. Learners are also more likely to receive feedback from an educator if they have already self-assessed their performance and know how their current actions impact a larger goal.

What does Principle 16 look like in action?

- In an electricity workshop: Educator builds excitement in the learners about seeing their lightbulbs light when creating a lighting board. Every time a student's light comes on, they give a standing ovation to the student then asks the student to complete a self-assessment worksheet on their own performance.

Principle 17: Learners learn better when they think about how they learn.

What is the best location for you to study? Do you prefer complete silence, or is background music more helpful to you? The ability to think about our own thinking is called metacognition. Learners with metacognitive abilities achieve at higher levels (McCormick, 2003). As learners increase awareness of how they acquire knowledge, learners can learn to regulate their behavior to optimize learning. They begin to see how their strengths and weaknesses affect how they perform and are more likely to play a role in their own self-assessment of knowledge (Zimmerman & Schunk, 2001). Metacognition can also play a role in learners becoming reflective about other areas of their lives. Researchers cite correlations between higher metacognitive abilities and increased self-awareness in social settings, increased ability to understand differences in others, and increased intrinsic motivation.

With so many benefits to metacognition, we might be wondering how educators can help learners improve their metacognitive abilities. Most importantly, educators can help learners recognize that their minds are programmed for growth. Teaching learners growth mindset concepts can increase their academic and social achievements (Dweck, 2006). Educators can also provide additional practice for concepts learners struggle with, prompt learners to reflect on coursework and their understanding of concepts, and increase the cognitive level of assessments, which allows learners to deepen their knowledge of a subject.

What does Principle 17 look like in action?

- In an agricultural biology class: Educator asks learners to complete an exit ticket which responds to three prompts: (1) What was the easiest concept for me to understand? Why? (2) What was the most challenging concept for me to understand? Why? (3) What strategies will work best for me when I study for the genetics exam?

Putting It All Together

Becoming a masterful educator is about more than just organizing curriculum and delivering instruction, it also requires an understanding of the core principles that help guide instructional decisions. Think back to your favorite educator. Which of the principles of teaching and learning could you identify in the way they taught? Incorporating all the principles in this chapter into daily instruction and action as an educator can take time. As we develop our skills as educators, these principles can help guide the way we think about teaching and learning. By taking time to align what we do to these principles, we are taking the first step toward becoming the educators that someone will reflect on one day as their favorite.

Still wondering how these lessons look in practice? The best way to become proficient at integrating these principles is to practice. Some practice tips for principles of teaching and learning in each area are as follows:

- **Principles of Organizing Content**
    - Practice planning out a scope and sequence for a course over a school year. Begin by determining the course objectives and then break down the things learners should know or do into a logical sequence of events.
    - Practice writing lessons that include all the nine events of instruction.
    - Have a peer, mentor educator, or educator education faculty review your unit plans and daily lesson plans. Their expertise can help you identify areas of strength and areas for improvement.
    - Teach your lessons! Sometimes the best way to find the errors in organizing content is to present your lessons to learners. Reflect at the end of the lesson and think about which principles were or were not embedded or shared with learners.
- **Principles of Delivering Instruction**
    - Spend some time thinking about your own progress toward each of the characteristics of an effective educator as set forth by Rosenshine and Furst. Which of the characteristics come naturally to you? Which will you need to spend some time focusing on integrating? This reflective process can allow you to pinpoint areas for continued improvement. Some of the best educators honed their skills through careful reflection and continual improvement, and you can too.
    - Video yourself teaching. Sometimes watching a video of your own instruction can be a powerful glimpse at what learners see when you are teaching. Take the time to identify places where you demonstrated principles of delivering instruction and make a note of principles that were not evident. As a bonus, you can see what university supervisors and administrators get to see when you are evaluated on your teaching abilities. Find the rubric they will use to evaluate you and complete your own self-assessment.
- **Principles of Student Learning**
    - Review your lesson plans and make sure you have taken into consideration differentiation in instruction, student motivation, how you will provide feedback, and how you will help learners learn about their learning.

- Take advantage of every opportunity to learn from each student how to be a better educator. Make a note of what works for learners and make every effort to replicate things that have worked in the past.

Learning Confirmation

1. Define the following terms:
    a. Teaching
    b. Learning
2. Explain why teachers should incorporate principles of teaching and learning in their instruction.
3. Describe the differences between Principles of Organizing Content, Principles of Delivering Instruction, and Principles of Student Learning.
4. Watch a recording of a teacher in action. Identify the principles of teaching and learning you can see exhibited in the lesson.
5. Make a plan for integrating each of the principles of teaching and learning within a lesson. How will you make sure to address each component?

Applying the Content

You are teaching a plant science lesson in a secondary-school-based agricultural program. Your lesson is on the parts of a plant and their functions. You know that your administrator will be coming in to observe your teaching and expects to see all the principles of teaching and learning on display. How will you develop a lesson that incorporates all the principles of teaching and learning? Outline how you will integrate the principles for planning instruction, delivering instruction, and facilitating student learning into this lesson.

Reflective Questions

1. Are there some principles that you feel more aligned with? Consider how you will use your personal strengths to use these principles to your advantage.
2. Are there some principles that you feel more disconnected from? Knowing how to integrate the principles you aren't as comfortable with can help you strengthen potential deficits in your teaching.
3. How do the principles of teaching and learning change based on the personalities of students or the teacher?

Glossary of Terms

- **teaching:** The process of mutual exchange and/or relationship among an educator, learners, and curriculum or society
- **learning:** The process of acquiring and transforming information into knowledge which can be assimilated, transformed, and processed by the learner

References

Darling-Hammond, L., & Bransford, J. (Eds.). (2007). *Preparing teachers for a changing world: What teachers should learn and be able to do*. John Wiley & Sons.

Dewey, J. (1938). *Experience and education*. New York: Simon and Schuster

Dweck, C. S. (2006). *Mindset: The new psychology of success*. Random house.

Gagne, R. M. (1965). *The conditions of learning and theory of instruction*. Holt, Rinehart and Winston.

Harlen, W., & James, M. (1997). Assessment and learning: differences and relationships between formative and summative assessment. *Assessment in Education Principles, Policy & Practice*, *4*(3), 365–379.

Keller, M. M., Hoy, A. W., Goetz, T., & Frenzel, A. C. (2016). Teacher enthusiasm: Reviewing and redefining a complex construct. *Educational Psychology Review*, *28*, 743–769.

Norman, D. A., & Spohrer, J. C. (1996). Learner-centered education. *Communications of the ACM*, *39*(4), 24–27.

McCormick, C. B. (2003). Metacognition and learning. *Handbook of psychology*, 79-102.

Rock, I. (1957). The role of repetition in associative learning. *The American journal of psychology*, *70*(2), 186–193.

Rosenshine, B., & Furst, N. (1971). Research on teacher performance criteria. In B.O. Smith (Ed.), *Research in Teacher Education* (pp. 37–72). Prentice Hall.

Rotter, J. B. (1966). Generalized expectancies for internal versus external control of reinforcement. *Psychological Monographs: General and Applied*, *80*(1), 1-28.

Sousa, D. A. (Ed.). (2010). *Mind, brain, & education: Neuroscience implications for the classroom*. Solution Tree Press.

Tinto, V. (1990). Principles of effective retention. *Journal of The First-Year Experience & Students in Transition*, *2*(1), 35–48.

Walker, H. M., & Buckley, N. K. (1968). The use of positive reinforcement in conditioning attending behavior. *Journal of Applied Behavior Analysis*, *1*(3), 245–250.

Zimmerman, B. J., & Schunk, D. H. (2001). Reflections on theories of self-regulated learning and academic achievement. *Self-regulated learning and academic achievement: Theoretical perspectives*, *2*, 289-307.

4. LEARNING AS PROBLEM SOLVING

Carla Jagger

Setting the Stage

We all face countless problems each and every day, whether it is deciding what to wear based on key factors about our day or trying to solve an issue at work. While these seem like rather insignificant problems, we still go through the same processes for these problems as we do for the more complex and challenging issues we inevitably face. When we look at our agriculture programs, we encounter and present problems constantly through our curriculum and even through the basic operations of our classroom. Let's consider the following situation together:

> *Your students start noticing that the classroom pet is not eating all their food and they have not needed to fill the pet's water as frequently either. They are growing concerned and are unsure of how to address the situation or determine the potential causes of the pet's new behavior.*

When students bring this issue forward, it might be easy for us to offer possible causes and answers to their questions based on our knowledge of the pet. However, more in depth learning can take place if we help facilitate and lead the students through an investigation of their own to determine possible solutions.

Objectives

Within teaching and learning, problem-solving is a key method for us to utilize so we can help our learners practice and master the process of solving complex issues as they arise. You will also see, as we unpack this method of teaching, that the learning process is placed on the students with our guidance, making it a student-centered approach to teaching. By the conclusion of the chapter, you should be able to:

- Define the problem-solving approach.
- Identify best practices for executing the problem-solving approach.
- Utilize problem-solving as a teaching strategy.

Introduction

Problem-solving is not a new concept in Career and Technical Education and agricultural education. Many individuals have contributed to our knowledge of problem-solving in teaching and learning within the agricultural setting over the decades. This list includes Krebs (1967), Crunkilton & Krebs (1982), Gubitz (1984), and Hedges (1996), many of whom link their work back to the foundations of problem-solving laid out by John Dewey.

Overview

If we look back at Dewey's foundational work, we see him place importance on the social nature of learning. He stressed a means of problem-solving that focused on the social role of people and stressed that our society benefits from the broader community and by making connections across cultures. Tanner (1997) evaluated Dewey's Laboratory School and the concepts that educators should still consider when developing our teaching and learning approach, which includes educating students in the problems of living together. In our agricultural education programs, we have a unique opportunity to connect our content back to issues in our local communities as well as statewide, nationwide, and worldwide. Additionally, there is an opportunity to foster service-involved thinking and social sensitivity and to create socially and civically responsible individuals through our programs. Problem-solving and other heuristic approaches (i.e., inquiry learning, discovery learning) to teaching and learning are a means by which we can achieve some of our internal goals of creating well-rounded, civically engaged students.

While there are many variations when defining problem-solving as a teaching approach, for this chapter, we will use Moore's (2015) working definition of problem-solving as "the intentional elimination of uncertainty through direct experiences and under supervision" (p. 353)—in other words, creating space for learners to generate new knowledge using problem-solving strategies with guidance. These problems can be planned and prepared for or spontaneous, occurring on their own. The key points of this definition are "direct experiences" and "under supervision." Problem-solving is an active teaching approach and is student-centered, which means students will be directly involved with the creation of knowledge regarding the problem at hand. Additionally, "under supervision" should not be forgotten as a part of this definition. Although students will encounter problems on their own to work through, we should still provide supervision, guidance, and advice in all situations that are part of our curriculum. As we begin to unpack the problem-solving approach, you will see there is a systematic approach to this teaching technique.

Executing the Problem-Solving Approach

Within our various teaching spaces in agricultural education, we will typically plan for problems and issues we want our students to solve. We do this through creating Units of Instruction and laying out problem areas within the content. Embedding problem areas within our Units of Instruction helps to bring focus to the content and should be connected to other areas of instruction to help with students' transfer of learning. If we do not properly identify the problem area, our approach may not be as effective. Without proper planning, we could have a problem that is too narrow, which will limit the analysis we can lead our students through, or we could plan for a problem that is too broad, which could result in confusion. Additionally, while planning our overall program of study, it is important to line up our Units of Instruction and problem areas using a seasonal approach based on your regional area and community needs (e.g., developing an integrated pest management plan around the specific plants being grown in your school's greenhouse, identifying how to protect against common diseases based on the crops currently growing in your area).

While we should plan for problem-solving, we also cannot discount the "teachable moments," the problems and issues that just naturally arise in our program that can be solved. It is important to involve your students in the generation of solutions and use the problem as a learning experience when possible. If we plan for the problem-solving approach and build our ability to execute this method, it will become natural for us to include students in those other issues that come up on their own (e.g., sick animals, equipment breaking, an immediate school or community need).

Steps to the Problem-Solving Approach

Hopefully by this point you are wondering how to execute this strategy within your teaching. The problem-solving approach includes five basic steps, which you'll see vary from source to source. In general, these include:

1. Identify the problem
2. Analyze the problem and gather information
3. Generate potential solutions
4. Select and test solutions
5. Evaluate the results

Identify the problem. One of the most challenging elements of problem-solving as a teaching approach is clearly defining the problem to be addressed. The problem itself should not be too narrow or too board as both can lead to confusion. Unless a problem naturally arises, we must plan for the problem and clearly present it to our learners. In addition to identifying the problem, we also need to create interest and motivate our learners to want to explore solutions to the problem. Motivating our learners is ultimately the hidden step to the problem-solving approach, but this substep should not go unaddressed. Refer to the Principles of Teaching and Learning related to motivation for good teaching actions for motivating your learners.

Analyze the problem and gather information. Once the problem has been clearly presented, students should move into the next step to explore the problem. As with most behaviors, you will probably need to spend a little time training your learners how to efficiently analyze the problem. Train them on how to track down credible sources, how to critically analyze information, how to look at all possible angles, and how to synthesize the information gathered, just to name a few approaches. Once you help with the collection and analysis of information a few times, you should be able to take a more hands-off approach with that same group of learners as you continue to facilitate more problem-based instruction. The analysis of the problem should help lead to the next step of generating possible solutions.

Generate potential solutions. During this step you should let your learners share all potential solutions no matter the expenses associated or how crazy the idea might sound. Encourage your students to justify their solution based on the information gathered to help keep them on track. Depending on how you design the activity, invite all learners to share potential solutions. For example, if you are solving the problem as a whole class make it an assignment for each student to generate their own solution(s). Or if you have small cooperative groups working on the problem, put in safeguards like assigning student group moderators who will ensure all students have a voice in the process (see the section on group teaching techniques in chapter 8 for more ideas). Do your best not to limit creative thought since you will begin to collectively identify the better solution(s) to test as you move into the fourth step.

Select and test solutions. As you can see this is a two-step action of the problem-solving approach. You will begin this step by evaluating the list of possible solutions based on feasibility as a class. Since students were not limited in their thinking during the previous step, you will need to have the conversation now about what you can test based on your resources. You could start by asking your students to identify any of the solutions that might be too extreme in terms of costs, time, resources, and so on. Once you have narrowed down your list, it will be easier to identify the better solutions to test. If you have multiple feasible and testable solutions, consider splitting up the class to test the possible solutions, or you can test them out together one-by-one as a whole class.

Evaluate the results. Once you have successfully tested your solution(s) have your students evaluate the results. Was the solution successful in addressing the problem? If so, how could the solution be improved to better address the problem? If it was not successful, how could you modify parts of the solution to test it again? These are just a few questions you could have your students address as they reflect on the solution(s) tested. As you know, we can learn a lot from failures during the problem-solving process so do not discount failed solutions, make them a learning opportunity. Even if your first tested solution is successful, have students evaluate the process to identify if a more effective and efficient approach could exist. At a minimum, make sure you evaluate the results as a whole class. If cooperative groups were used during the process, consider having each group share their results so everyone can learn from one another.

Good Practices while Executing the Problem-Solving Approach

Earlier you were introduced to the Principles of Teaching and Learning in chapter 3; you'll therefore be able to identify several of these principles through the good practices laid out below. In other words, as you incorporate the Principles of Teaching and Learning into your lessons you will also be using these good practices and vice versa. While utilizing problem-solving techniques, there are several actions that we can incorporate which will help make our students' learning meaningful (Krebs, 1967).

1. Help students connect the parts of the problem situation to the broader issue. Typically, our students hold novice knowledge on topics we cover. As novice learners, students need guidance in making connections between topics and identifying larger issues that may play a role in the problem.
2. Incorporate the immediate use of knowledge. We know this is an important step for all learning; if immediate action is not taken with a problem situation, students may fail to see the problem's relevance and importance.
3. Make the objectives of the problem-solving approach clear. There are several approaches described later in this chapter that can be used when laying out a problem situation for students. All the described techniques will help give the problem situation structure and clarity when presenting the problem to your learners.
4. Build in transfer of knowledge connections. Each problem situation we present will typically be very specific, so we should aim at building in additional opportunities for students to use the problem-solving techniques in other situations.
5. Use a variety of activities within the learning situation. Using a variety of learning activities can reach learners in different ways. With the steps of problem-solving laid out earlier, we can utilize a variety of individual and group teaching techniques to help accomplish the objectives of the problem-solving approach.
6. Connect the problem situation to the students' personal lives to make the learning more meaningful. Additionally, build in opportunities for your students to set their own problems and inquiry ideas.
7. Incorporate cooperative learning. We learn from each other with problem-solving so work on making the activity a collective effort.
8. Allow students to succeed. We know that success is a strong motivating force. Students who experience failure often will tend to shut down in the learning environment. When we start to practice problem-solving with students, we should build in opportunities for our students to be successful in discovering solutions as we train them through the approach.

Pitfalls to Avoid when Executing the Problem-Solving Approach

In addition to the good practices above, there are several pitfalls we should avoid with a problem-solving teaching approach. As we reflect on each problem-solving teaching experience, we begin to recognize these pitfalls that lead to confusion and frustration for both us and our students. Naming these pitfalls in advance and building in supports to avoid them will help with the effectiveness and efficiency of the problem-solving teaching strategy.

1. Unsuitable unit or problem area. Make sure the unit and problem has relevance to the students and your community. This also ties back to taking a seasonal approach to problem-based learning.
2. Inadequate teacher analysis of the situation related to the problem area. When planning for the problem situations, we should do our own research of the problem and have a general direction in mind for possible solutions before the students start walking through the steps of problem-solving on their own.
3. Inadequate teacher objectives. Not introducing appropriate objectives can lead to confusion of the problem situation and execution of the techniques you have tried to plan.
4. Omission of the interest approach or allowing the interest approach to become the problem-solving situation. Omitting the interest approach may lead to a lack of student motivation in solving the issue. We know that learning is not passive; without motivation there will be no drive for the students to participate in the learning.
5. Poor discussion-leading technique. We should use appropriate prompting and probing questions during discussion to help facilitate learning while also giving students creative freedom. If we are not careful, we could be too leading in our prompts or completely give away potential solutions which tends to defeat the purpose of the problem-solving approach.
6. Lack of orientation at the start and the end of the class period. As with any content or approach to teaching we need to include appropriate lesson introductions and summaries that can help students see the larger picture and make cognitive connections.

Overall, there is a lot of variation you can add to all these steps when executing the problem-solving approach. Keeping the basic steps in mind, along with the best practices and pitfalls to avoid, will help to provide overall structure to the activity. In addition to problem-solving, there are several other heuristic teaching strategies we can incorporate in our teaching, including inquiry-based instruction, discovery learning, and experimentation.

Other Heuristic Teaching Strategies

While these approaches are very similar to problem-based instruction, there are differences in the purpose and structure. Moore (2015) lays out each of these additional methods including discovery, inquiry, and project-based learning methods. Discovery learning in essence is a formal approach to problem-solving using the scientific method. The steps involved with discovery learning are nearly identical to the five steps already established for problem-solving. With this approach you would identify a problem, develop possible solutions, collect data, analyze and interpret data, and test conclusions. As you work through this strategy, the steps will ultimately lead to the need for revised solutions or provide supporting evidence for the chosen solution.

Inquiry learning is another problem-solving strategy, where emphasis is typically placed on the process of determining a solution rather than finding a solution for the problem. This approach would be very useful when trying to work through more of a hypothetical problem when you may not have all the resources needed to test possible solutions. There are multiple approaches to inquiry-based instruction (Moore, 2015) including a three-step procedure (identify the problem, work toward solutions, and establish solutions) and the five-E inquiry sequence (engage, explore, explain, elaborate, and evaluate).

Additionally, project-based instruction, although not always tied to a problem situation, can at times have a natural connection to the problem-solving depending on your approach to student engagement with the problem. For instance, you may decide on a specific product the students need to create as they determine solutions such as case studies, surveys, models, and so on. With all these strategies, you typically have options on how structured or flexible you want to make the active learning for your students.

Encouraging Individual Problem-Solving

In addition to working through problem situations collectively, our learners are also engaged in problem-solving through their Supervised Agricultural Experiences (SAE), in which they will need to solve problems on their own. Problem-solving is a skill which needs to be developed and practiced like any other skill, adding to the value of us incorporating problem-solving to our classroom learning so students can solve their own problems successfully. Gubitz (1984) identified several questions that can be proposed for learners to engage in problem-solving on their own, including:

1. What is the specific problem?
2. What really is the source of irritation?
3. Why is it so bothersome?
4. Who is involved?
5. When did it start?
6. How is the problem "ours" to do something about?

While these questions are somewhat broad, we can tailor them for our students and their situations. We know that students will face many problem situations while carrying out their SAE projects and it is important that we help facilitate their identification of solutions instead of just providing them with a solution. Most of us have the goal of learners becoming independent and being able to establish knowledge on their own, but this ability comes with practice. Several methods for providing practice to learners are laid out in the following section.

How Problem-Solving Looks in Agricultural Education

There are many ways to set up a problem-solving situation for your students. Hedges (1996) created a resource for Career and Technical Education (CTE) that lays out several of these approaches along with lesson vignettes from in-service CTE teachers. All the example approaches explained here are summarized versions of the original work laid out by Hedges and the contributing teachers (1996).

The first technique for laying out a problem is **key steps**, used when there are specific steps required for the successful completion of a task. Typically, the Key Steps technique will be partnered with a demonstration of the needed steps. This technique is extremely useful for objectives centered around using and maintaining equipment. No matter which curriculum we are using, nearly all courses have equipment of some kind that require the set of steps to use it properly.

The second technique is the **forked-road** approach, where there are two choices for the problem solution, and you need to consider the advantages and disadvantages of both options. Looking at an example from animal sciences, you may choose to use this technique when asking students to decide when to keep or cull a cow from the herd, or from veterinary science, when deciding to use euthanasia in determining whether treatment is possible with a sick animal. When carrying out this technique you want to make sure you identify the decision that needs to be made as well as lay out all key factors that should be considered in making an informed decision. As students reach their decision between the two options, it is good practice to have each student justify their choice.

The **possibilities-factors** technique sets up a problem with more than one solution when several factors must be considered to select the most appropriate solution. You might plan for this approach when asking students to choose the most appropriate harvesting or transplanting method for a specific plant. For this technique, you will want to clearly present the problem statement and have your learners list all the possible choices (this step is also great for review of previously learned content). Once all the possibilities have been listed, you would ask the students to discuss all the factors that should be considered before making their decision.

The **situation-to-be-improved** approach can be used when specific characteristics or requirements should be considered for a situation. Students are presented with important information about the situation such as the requirements needed for the successful execution of the situation, and they would then give recommendations for improvements. This technique could be used in a natural resources unit when examining the interdependence of an ecosystem. For example, you might present students with the issue that a local landfill will reach its maximum capacity soon, so how can you make it last longer? After the issue is presented, the students would then discuss characteristics to be considered and possibly layout the "ideal" situation. Once that information is identified, they would determine the *what* and *why* for all characteristics to help them arrive at recommendations for improvement of the situation.

A fifth approach is the **effect-cause** technique, where causes of an issue would be evaluated to determine appropriate options for action. For this technique, we can consider an example from landscape or turfgrass management: a mower is not cutting evenly when used on a putting green, what is the cause of the problem? To help students determine the cause, additional information would need to be presented; this information could potentially be presented in a case study format or given more simply, such as by adding that the mower leaves an incline with every pass and the operator has already checked the air in the tires as well as the height of the deck and both were fine. These additional details will help students make a more informed decision when trying to determine the cause of the issue.

Finally, problems could be presented using the **four-question interest approach**. Four questions should all be considered by the learners:

1. How important is ...?
2. What problems have we had with ...?
3. What do we need to know or be able to do to correct or prevent these problems?
4. What related information are we lacking?

Using this technique can help motivate students by involving them in the planning and execution of the lesson. Through their responses to the four questions, as an interest approach, they will develop ownership for the direction and success of the lesson and provide a foundation for critical thinking and selecting which additional problem-solving techniques should be used. An example of this technique would be, if you had a group that has limited knowledge of identifying and interpreting nonverbal messages, the specific questions you would ask the learners would then be: How important is it to be able to recognize and interpret nonverbal communication? What problems are associated with reading nonverbal messages? And what do we need to know and/or be able to do in order to solve and/or prevent these problems? The students would provide answers to all these questions within the interest approach of the lesson, and then you would be able to use additional problem-solving techniques as you move through the actual lesson content. The fourth question—What specific information are we lacking concerning what we said we need to know and/or be able to do?—is more of a lead-in question to begin teaching around the issue or specific skill.

Learning Confirmation

1. Identify a problem situation you will present to your learners and layout which problem-solving technique you will use for the situation. Make sure you name the technique and establish how you will facilitate each step of the problem-solving approach.
2. Using a Venn diagram, compare and contrast two problem-solving techniques of your choice and consider the characteristics of the technique and when each technique would be used and why.

Applying the Content

1. Prepare or adapt a technical lesson that can be taught in a local context utilizing one of the problem-solving techniques.
2. Utilize one of the problem-solving techniques to create an interest approach for an existing lesson.
3. Use one of the problem-solving approaches to create a nonformal lesson with a leadership topic.

Reflective Questions

1. What are the five steps of the problem-solving approach?
2. What are similar teaching techniques to problem-solving?
3. Which problem-solving technique would be best when trying to make a decision between two options?
4. What specific supports will you put into place when designing a problem-solving activity to conduct with your learners?

Glossary of Terms

- **problem-solving:** As a teaching approach, "the intentional elimination of uncertainty through direct experiences and under supervision" (Moore, 2015, p. 353)
- **key steps**: Technique to problem-solving used when there are specific steps required for the successful completion of a task
- **forked-road:** Problem-solving approach where there are two choices for the problem solution, and you need to consider the advantages and disadvantages of both options
- **possibilities-factors:** Problem-solving technique that sets up a problem with more than one solution when several factors must be considered to select the most appropriate solution
- **situation-to-be-improved:** Problem-solving approach that can be used when specific characteristics or requirements should be considered for a situation
- **effect-cause:** Problem-solving technique where causes of an issue would be evaluated to determine appropriate options for action
- **four-question interest approach**: Problem-solving technique composed of the following four questions: How important is ...?, What problems have we had with ...?, What do we need to know or be able to do, to correct or prevent these problems?, and What related information are we lacking?

References

Crunkilton, J. R., & Krebs, A. H. (1982). *Teaching agriculture through problem solving* (3rd ed.). The Interstate Printers & Publishers, Inc.

Gubitz, A. C. (1984). *Instant personal problem solving: Simplified approaches for everyone*. The Interstate Printers & Publishers, Inc.

Hedges, L. E. (1996). *Teaching for connection: Critical thinking skills, problem solving, and academic and occupational competencies*. Ohio Agricultural Education Curriculum Materials Service.

Krebs, A. H. (1967). *For more effective teaching: A problem-solving approach for teachers of vocational agriculture* (2nd ed.). The Interstate Printers & Publishers, Inc.

Moore, K. D. (2015). *Effective instructional strategies: From theory to practice* (4th ed.). SAGE Publications.

Tanner, L. N. (1997). *Dewey's laboratory school: Lessons for today*. Teachers College Press.

5. INCLUSIVE TEACHING

Stacy K. Vincent and Donna Westfall-Rudd

Setting the Stage

In the early 1970s, Dr. Henry Schmitt asked, "Will middle-class White Anglo-Saxon Protestant vocational agriculture teachers accept minority youth and adults enrolled in vocational education in agriculture?" (1971). Schmitt posed this question to agricultural education institutions soon after the desegregation of public schools, the creation of consolidated school districts, and the merger of the New Farmers of America (NFA) and the Future Farmers of America (FFA). His corresponding study focused on designing "professional experiences that will prepare teachers for urban and rural minority children, youth, and adults" (p. 20). The chapter is grounded in the multicultural education literature (Banks, 2008) that coincidentally began to emerge at the same time Dr. Schmitt published his dissertation. Since Schmitt's study, secondary and postsecondary agricultural educators have had sporadic discussions about the learning needs and experiences of underrepresented minority students in local programs. Still today, we, as educators, face the challenges of including all students in local agricultural education experiences.

Objectives

The purpose of this chapter is to provide readers with the opportunity to:

- Discover the importance of developing knowledge of individual cultures and identities to support and engage students in agricultural education.
- Investigate how to connect intersecting student and teacher identities within agricultural education courses to obtain multicultural autonomy.
- Apply multicultural and inclusive teaching practices in an agricultural education classroom.

Introduction

Unfortunately, society constructs an identity about an individual without ever getting to know them. For some individuals, this may be a positive construction of identity; however, it may never reflect who one truly is, nor how they want to be known. **Social identifiers**, as listed in table 5.1, are dispositions that are integrated into one's identity and therefore dictate their interactions with other individuals. On the other hand, **personal identifiers** are distinguished characteristics for which the individual personally recognizes who they are and whom they associate with based on commonalities.

Overview

Addressing the needs of all students can seem like a daunting task for early-career educators. This chapter is intended to help educators begin the process of purposefully planning courses, FFA opportunities, and Supervised Agricultural Experiences (SAEs) to be inclusive of all student cultures and identities that may be present in the local program. While this may initially seem too big of a challenge, this chapter intends to offer a foundation of knowledge and recommended practices to support the implementation of multicultural and inclusive teaching practices.

Social Identifiers	Personal Identifiers
Race/Ethnicity	What Makes You Unique
Language	Talents
Gender Identity	Likes
Geographic Location	Peculiarities
Socioeconomic Status	Ways of Doing Things
Religion	Personality Type
Age	Introvert/Extrovert
Sexual Orientation	Skills
Education	Passions/Compassions
Body Type	
Ability	
Family Structure	

Table 5.1: Social and personal identifiers.

Multicultural Autonomy

Are we good teachers? Of course, we want to be good teachers and desire to make a positive difference for students. But what does it take to make a positive difference in the lives of students? Though perhaps surprising to our noneducation friends, we would all agree that simply providing the details for learning a particular task does not reflect a methodology that inherently creates a positive life-long connection.

When examining our reason for considering the profession of educating others about agriculture, our inspiration comes from a myriad of approaches, such as respect toward a previous instructor, a passion for helping youth, or our deep inquiry within the content area. Coincidentally, whichever approach, our vision of who we will deliver the content to is skewed by what we know. To simplify this understanding, let's examine your palette. Let's say that your entire life, the only meal you had consumed and visually recognized was pizza; how could you envision, understand, and relate to the taste, smell, and visual of sushi? It cannot happen. The same visualization for our attempt to educate others is biased toward only what we have seen, experienced, and exposed ourselves to.

This process is called the Apprenticeship of Observation (Lortie, 1975). We gain efficacy in our ability (in this context, teaching) based on our observations and experiences in the past. Therefore, it is difficult to identify what our classroom, and who the students we will educate, will become if we have never exposed ourselves to anything different. While the diversification of exposure does not predict your ability to be a quality educator, it limits the overall volume and diversification of whom you positively impact.

But what if we were knowledgeable in the pizza community as well as the sushi community? And what if the more we expand in the pizza and sushi community, our awareness, knowledge, and attitude grow and expand? Now we have the ability to connect the two communities so they can find ways to work together. Similarly, a teacher can gain students' trust from various cultural backgrounds without letting go of their own identity. In doing so, we must immerse ourselves in the different cultural communities to expand our competence levels. When an individual can competently work and communicate with multiple cultural groups and be accepted within the groups, we call this ability multicultural autonomy (Vincent & Westfall-Rudd, 2022). The individual would not have to change their core identity to fully obtain multicultural autonomy. An individual cannot simply wake up and determine that they have multicultural autonomy. This concept takes work and patience and is ongoing. To reach the point that autonomy is possible, a teacher must begin with self-reflection of (*a*) awareness, (*b*) attitude, and (*c*) skills. For a visual of multicultural autonomy, see figure 5.1.

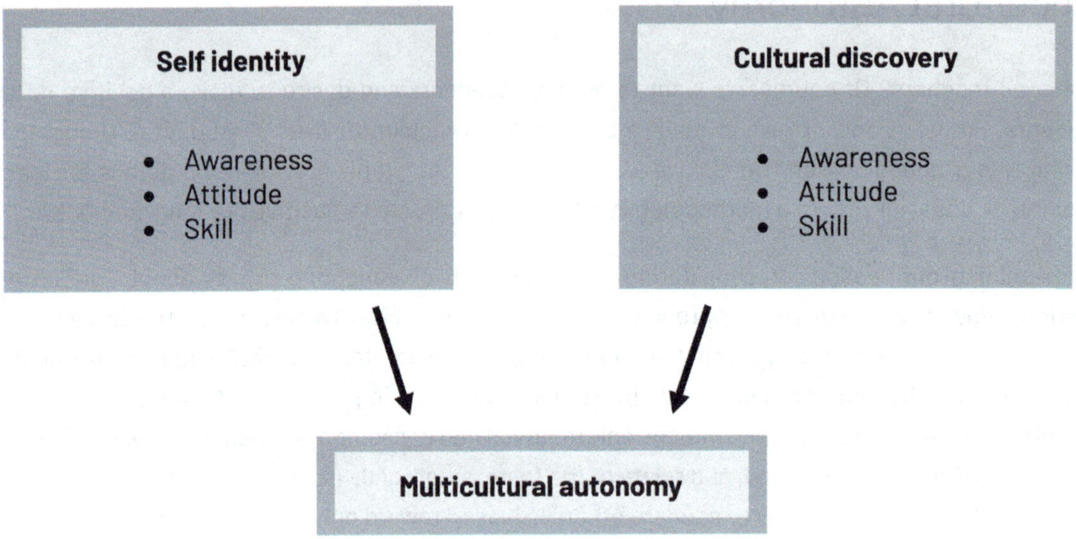

Figure 5.1: Multicultural autonomy growth model.

Self-Awareness

Harris (2001) recognized that a lot goes into the identity of self. We are labeled and identify ourselves, whether it is correct or not, by *social identity* (i.e., race/ethnicity, language/dialect, religion, age, sexual orientation, educational level, body type, socioeconomics, ability, family structure, geographic location, etc.). Social identifiers can be helpful and harmful in the perceptions of those around us. One of this chapter's authors, Stacy, recalls a time social identifiers impacted him:

> *"I recall, as a child, on a weekend trip out of my home state, my parents sent me to get extra towels from the hotel staff. The hostess loved my accent, and when she discovered I was from Kentucky, she looked over the counter and said, "Wow, you have shoes!" My social identity, in this instance, personified a negative stereotype."*

However, our self-identity also encapsulates an area many individuals never see or value if it is not recognized in our workplace, also known as *personal identity*. Our identity (i.e., talents, likes, peculiarities, personality, political beliefs, ways of doing things, introversion/extroversion, skills, uniqueness, etc.) also plays a major role in defining who we are.

Before we can ever have more than a minimal impact in our classroom, we must become aware of who we are. While our lives continue to experience memories and our interactions diversify, so should our continued reflection and awareness, which inadvertently changes our self-identity.

As teenagers, we grew in our self-awareness, but it didn't keep adults from telling us to check our attitude. It is good advice (although probably not in the tone or format they were referring to). Gay and Kirkland (2003) believe it is crucial to self-reflect on our attitude during critical cultural moments. These critical moments often occur in the presence of major global events that may be happening. Although we may ignore many of the issues, we must take a second to reflect on our attitude as to why we are choosing to ignore it. The same reflection must occur when we are experiencing anger, sadness, excitement, happiness, and disappointment (Vincent & Drape, 2019). It is very important to note that the reflection of these attitudes is explicit and should not be confused with implicit attitudes (Benaji & Greenwald, 2013). Still, this chapter will not be expanding into it.

Self-Attitude

As Stacy was entering high school, the rapper (at that time, now better known as an actor) Ice Cube, released a song titled "Check Yo Self," and one of the lines that always stuck with him in the song was *So come on and chickity check yo self before you wreck yo self.* Cube (as Stacy refers to him, because they're buds) understood self-identify's second construct: self-attitude. In this area, we begin to examine our attitudes.

It is not difficult to determine our attitude by simply taking a stroll through our social media outlets and examining what we like, share, post, read, and follow when a topic of controversy comes across our feed. Now, we may not follow a lot on social media, but we do speak with colleagues on occasional topics or read articles regarding the mishandling of particular issues. In these instances, do you ever find yourself reading the comments that some brave souls posted? During your read, you will find yourself shocked by the ludicrous statements or nodding your head to the realization that someone else believes the same thing. At that instant, we should begin to restate, reflect upon, or simply rap Ice Cube's lyrics and simply "check yo self."

Attitude is crucial in how others perceive us and how we interpret someone else's attitude. Of course, we all have our peeves, but by not keeping your attitude in check, you begin to "wreck yo self." Let's go through an example together.

Silvia is having a difficult day on her way to school. There are relationship issues between her and her boyfriend, and her mother just told her that she could not stay after school due to parental assistance needed once she returns home to her adolescent siblings. At the same time, you are having a difficult day as well. You arrived a few minutes late and spilled coffee on your pants a few minutes before class began. In expressing herself only as she understands, Silvia walks into your classroom, slams her books on her desk, and storms out, muttering explicit words (also from an Ice Cube song).

How do you react? Do you go after her, express your feelings toward her methods, and ask that she share her feelings with the school principal? Do you take a second to realize that you (*a*) have a coffee-stained pair of pants and (*b*) are not in the right attitude to come to grips with how her day is crumbling? Most of us have seen how option one plays out. The teacher and the student go into an embarrassing shouting match where the student, the student's guardian, and the student's siblings will have a very bad day, and you end up on a video blasted on social media. Did you win? Maybe you got a shot in the ego arm, but you gained nothing in your ability to examine your self-attitude and continued to make impulse decisions no matter the spirit of your attitude. Honestly, attitude can be an endangering or a strengthening tool.

Self-Skill

Before having a classroom of our own, we determine the teaching style that works best for us and the students we are teaching. How do we know, though? Of course, our peers tell us it was a great lesson, but they have had similar classes from the same professors and received the information exactly like you did. The students to whom we provided the lessons may have seemed engaged, but were they from the best practicum of classes with students who are passionate about the content? Even our professor could inform us that it was structurally a sound lesson, but they are now removed from the culture in the classroom that you are about to enter. Bottom line, the group that we have provided lessons to thus far reflects very similar awareness levels, attitudes, and cultural beliefs to those that we have; therefore, we have to ask ourselves "does our pedagogical skill work in a classroom of students who think, look, and behave differently than those coming from the cultures to which we have ourselves been exposed?" Sometimes, we never fathom the extent of this question until we enter our classroom.

Creating a Multicultural Agricultural Classroom

There are multiple approaches that a teacher considers as they craft their lessons for the upcoming school year. One of the most important approaches is developing a curriculum that encompasses a multicultural path. Banks (2008) introduced four levels to approaching a multicultural curriculum (see figure 5.2) that can provide a scaffolding on which teachers can build their pedagogical skills and knowledge when creating a multicultural classroom and agricultural education program.

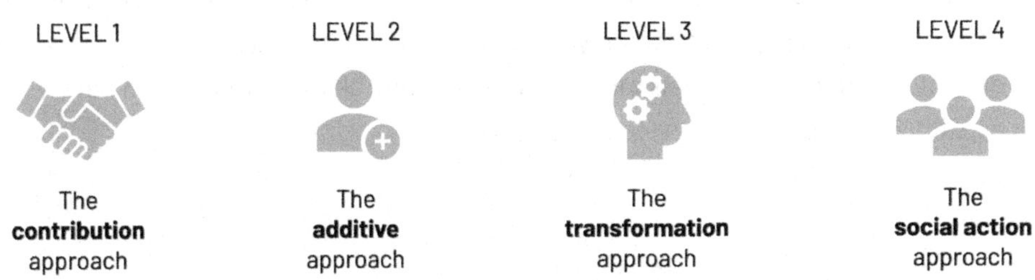

Figure 5.2: Approaches to multicultural curriculum reform.

Throughout the country, schools and classrooms are finding methods to integrate cultural differences inside and outside the community. One of the approaches, level 1, the contribution approach, occurs when teachers integrate holidays and cultural celebrations into the course activities, such as African American History Month, Cinco de Mayo, Women's History Week, and Asian American and Pacific Islander Heritage Month. The contribution approach is an initial step that compliments the other three strategies, primarily focusing on heroes, holidays, and discrete cultural elements. In the agricultural education curriculum, teachers can incorporate these cultural items into many of the courses in a local program. (See table 5.2.)

Subject	Level 1 Implementation
Floriculture	Introduce cultural holidays and images of floral designs seen in cultural holidays.
Animal Science	Identify the geographic origins of livestock breeds.
Food Science	Develop a bulletin board displaying the production and processing of foods specifically for various cultural markets to meet the diversity of cultural food traditions of holidays.

Table 5.2: Contribution approaches in agricultural education.

Educators are also beginning to frequently use practices that reflect level 2, the additive approach (table 5.3). At this level, teachers add cultural content, concepts, and perspectives without implementing structural changes or adding a large amount of effort to the curriculum revision. Banks says, "The additive approach is often accomplished by adding a book, a unit, or a course to the curriculum without changing the framework" (p. 47). Although level 1 and level 2 are approaches that compliment effort, they do not challenge the curriculum, the cultural norms, or the thoughts or interpretation of the dominant culture.

Subject	Level 2 Implementation
Introduction to Agriculture	• Integrate the history of agricultural practices of Native American cultures, including the origin of crop rotation and intercropping practices. • Integrate the history and movement of corn by Native communities from what is now called Central to North America. • Identify the history of crops originating in West Africa, including watermelon and peanuts. Break down the assumptions that only European settlers had the knowledge and skills to begin farming in North America.
Animal Science	• Have a discussion of enslaved West Africans chosen based on their animal husbandry experiences. • Have a discussion of animal husbandry practices from West Africa.
Food Science	• Examine the cuisine as it relates to global topography and geography—for example, the proximity of spicy food to the Equator. • Discuss how new food processing and preparation approaches were brought to the United States from West Africa. Dishes coined as Southern culinary cuisine, such as fried chicken, jambalaya, and barbecue, were first prepared by enslaved people. • Discuss how delicacies differ across the United States and the world. For example, "Cui," or a Guinea pig, is a meat served on special occasions in Ecuador. • Examine the food processing requirements of different religions.
Leadership Development	• Encourage students to present speeches in their first language.

Table 5.3: Additive approaches in agricultural education.

Level 3 fundamentally differs from the first two levels as it contains objectives that shift paradigms and question basic assumptions. By doing so, the curriculum allows students to see, hear, and comprehend different points of view. The level 3 transformation approach entails a structural change in the curriculum so that students are enabled to understand information from the perspective of diverse groups. Specifically, the aim of level 3 is to assist students in thinking critically and develop a skill set that empowers them to construct, document, and support their conclusions and generalizations. Table 5.4 provides examples of level 3 interaction in agricultural education.

Subject	Level 3 Implementation
Landscape Architecture	Examine socioeconomic opinions by discussing "Which individuals desire a landscaped home?" Taking a community trip and visiting various neighborhoods of different socioeconomic levels allows students to see that similar care is put forth, but that financial earnings play a role in what is on display. Nevertheless, some plants, regardless of income level, have value or meaning to the owner; thus, landscaping is important to all income levels.
Animal Science	When exploring countries of origin, lead students to determining why an animal was brought to the United States. Which ethnic group decided the animal that would inhabit America? What diseases did it bring to the land and whom did these diseases affect? Which religion supported the movement of these animals?
Floriculture	During a floral identification lesson, engage students in identifying the country of origin for each of the major flowers sold in the United States. Use research and discussion to encourage students to determine the cost of each flower sold during its peak season and what an individual is paid to grow and harvest the flowers in the country of origin (e.g., roses grown in Ecuador).

Table 5.4: Transformation approaches to agricultural education.

NOTE: As the curriculum moves into the transformation phase, the lesson becomes embedded and is fluid with what is being taught rather than a hard break to the curriculum.

When students begin to make decisions on important social issues by responding with action to help solve the problem, then the teacher has met level 4, the social action approach. We have the perfect opportunity for social action within agricultural education as our students are responsible for obtaining and managing an SAE project. In addition, agricultural education youth have the potential to be enrolled in our classes for multiple years, where social action can be built upon as students grow in their civic responsibility within their community. When the agriculture curriculum is being delivered in the social action approach (table 5.5), learning occurs, and proud moments are celebrated.

Subject	Level 4 Implementation
General to the program	Students develop a comprehensive list of agricultural magazines from various ethnic backgrounds for the school library to purchase and display.
Landscape Architecture	Students develop an arboretum that entails plant information that engages readers to think about the pain and suffering that occurred for the plant to migrate from its country of origin to the United States.
Agricultural Business	Student create projects focused on investigating the impacts of farm labor laws and policies on the immigrant and nondocumented agricultural workforce.
Leadership Development	Because of the diversity of the membership, the chapter officers develop an end-of-the-year celebration or banquet that is inclusive of all religious and cultural considerations, taking care to consider the day of the week, avoid conflicts with all holidays, and ensure all are able to enjoy. Students choose public speaking topics that respond to social justice issues impacting members of the community and the agricultural industry of the region.
Agriculture Sales	Students work with the local farmer's market so that the farmers are able to receive SNAP funding so low-income families can obtain fresh food.

Table 5.5: Social action approaches to agricultural education.

Teaching Practices for Multicultural and Inclusive Agricultural Education Classrooms

Changing teaching practices to create inclusive teaching strategies requires us to reflect on our current practices and classroom environment intentionally. Then, considering what we see and understand about our teaching and the spaces in which we teach, we can try new ideas. Once implemented, we must be courageous and ask our students for feedback on our changes.

Planning

- Review all IEP and 504 plans early to embed student needs within the curriculum rather than making targeted accommodations that can "single out" students.
- Identify English Language Learner students and their level of proficiency. Place them near the front of the room so targeted communication can occur.
- Designate a "Food Drawer" for students whose hunger needs are not being met.
- Develop a predetermined seating chart to assist and encourage student interaction and engagement rather than targeting disciplinary control.
- Obtain flags from the country of origin of your students and display them throughout the classroom.
- Develop bulletin boards that display the interests of every student and deliver a message that the classroom embraces a *SafeSpace* environment.
- Identify religious holidays during the school year and add them to your calendar as a reminder to bring attention to them.

As teachers, we all know that what things we say and how we say them matters. Students see us as leaders and role models in our classrooms. Therefore, to have positive and appropriate communication in our classrooms, we must ensure we do our part to make it happen.

Communicating with Students

- As part of activities on the first day of class, ask students to respond to the following questions on a worksheet you provide or an electronic format appropriate to the course:
    - What name would you prefer we call you?
    - What is something about yourself that you are proud of?
    - What would you like me to know about you?
    - What concerns you most about this class?
    - What is your favorite music genre?
- Always use a student's preferred pronouns and apologize when you misspeak.
- Post your preferred pronouns.
- Ensure classroom images and resources include representation of diverse cultures, races, ethnicities, genders, and abilities.
- Learn the cultural communication practices of the students in your program.
- Engage students in cocreating the classroom expectations.
- Ensure materials are ADA compliant.

Building relationships with students and their families through home visits has been a tradition in agricultural education. Over time, in many communities, teacher visits for SAE project supervision have been diminished or removed. As most of us agree, this is not a desirable change. We recommend renewing our efforts to make home or job site visits to strengthen the connections between our students' families and our agricultural education programs.

Building Relationships with Families

- Be an advocate for the student and bring parent requests to the attention of committee members during a 504 planning meeting.
- Constitute student praise calls rather than calls regarding demerits.
- Develop a bilingual monthly email to parents that discusses previous successes and upcoming events.
- Invite families to events and trips.
- Obtain funding for meal function events that alleviates the cost of attending (e.g., banquet tickets).
- Provide childcare opportunities for parents who desire to meet with you or attend a school event for your student.

Learning Confirmation

What would you do? Work with a colleague or a team of colleagues to discuss the following two case studies. How do you respond to the questions? Remember to provide a discussion that allows everyone to speak openly and be sure that to implement listening skills.

Case Study #1

A student wants to join the livestock team, has worked intently in class, and is passionate about the subject. They find out that practices are held after school, and if practices are missed more than once, it would result in the removal from the team. Unfortunately, the student cannot attend most practices because of their responsibilities. Living with only one parent, they must ride the bus with their younger siblings and watch them until their mom arrives.

Behind the scenes, the teacher is known to only allow teams of two males and two females OR four students all of one gender due to a philosophy of money "wasted" on hotel rooms that would only hold one person. Unfortunately, there are already three students of the opposite gender on the team, and while the student's talent is just as good, if not better than the current team members, they choose not to inform the teacher of their interest.

1. Does the student have a chance of making the team?
2. What amendments can be made to assist the (*a*) teacher, (*b*) student, (*c*) student's family, and (*d*) team members?

Case Study #2

A student (male, Muslim, and visually colorblind student) at the school has a very hectic schedule and enrolled in a floriculture course after hearing that it was a good class for students who needed a break from bookwork and desired a course where you could work with their hands.

It is the middle of the Ramadan holiday, and the subject is learning color patterns. On this particular day, the color wheel was the focus of the learning objectives, and the teacher began by saying, "All right ladies, using icing, food coloring, and sugar cookies; whoever correctly displays the color wheel will receive a pass on the next daily quiz." Unfortunately, the bottles of food coloring were not labeled, and once the student was midway through the assignment, the teacher approached and said, "Are you even trying?"

Once the class is finished and the assignments are graded, the teacher allows the students to eat their colorful cookies. The student offered their cookie to someone else and the teacher asked why they would give away something so delicious. Once the student tells the teacher their belief, the teacher simply says, "Oh, one cookie won't hurt you."

1. What were the three comments that could have created a microaggression?
2. How could the three comments have been avoided?
3. How will you ensure that you, and your classroom, are accepting of a student in a similar situation?

Applying the Content

Use the following questions to guide your development in establishing a more inclusive agricultural education program. Check them off when you have completed.

1. Do you know which students have an IEP and/or 504 plan?
2. Are your members comfortable attending an overnight trip with a student who identifies themselves in the LGBTQ+ community?
3. Does the program have an activity to participate in the community during Martin Luther King Jr. Day?
4. Does the officer team reflect the demographics of the school and agriculture education program?
5. Would your community be supportive of the program if the chapter president was an ethnic minority?
6. Do quiet students have a place in the program?
7. Would a vegan student feel comfortable talking to you?
8. Do you have a plan to help students who cannot read?
9. What will be the out-of-pocket cost for students to attend your most expensive trip?
10. Do you have alternative seating in your classroom for an obese student? Where will it be located?
11. What percent of your events will be outside the school day?
12. What percent of your students rely on bus transportation?
13. Do you have a liaison to assist you in the pronunciation of your students' names?
14. What music will you allow to be played in your classroom?
15. Have you set your calendar to recognize international holidays?
16. Have you created a "wish list" of places to visit, or list of places already attended, on every continent? Do you already know where to add your experience or "wish list" to the curriculum being taught?
17. Do you have a safe and secure area for students to maintain an SAE?
18. Can you identify a student who needs shoes and find the resources to provide?
19. Besides the school community, do you know of ethnic restaurants to expose your students to during a school trip?

Reflective Questions

1. How do you describe your individual identity and cultural background?
2. How do you use your identity and background to engage students in your classroom?
3. What multicultural and inclusive teaching practices will you implement in your classroom this semester?

Glossary of Terms

- **additive approach:** At this level of changing the curriculum to be more inclusive of all students, teachers adds cultural content, concepts, and perspectives without structural changes and with minimal effort to the curriculum revision
- **apprenticeship of observation:** One gains efficacy in their ability to teach based only on observations and experiences in the past as students, not on training received or practical teaching experience
- **contribution approach:** Occurs when teachers integrate cultural traditions, such as holidays, celebrations, and traditions, into the course activities, physical classroom visuals, or attempting to recognize through a sensory such as music playing on a particular day
- **culture:** All the complex aspects of a person or group's world experiences, including food, traditions, beliefs, habits of social interaction, environment, personal and social identity, music, and language
- **inclusive teaching practice:** Purposeful planning and implementation of teaching strategies that engage all students in the learning process regardless of ability, background, individual identity, or social status
- **multicultural autonomy:** When an individual can competently work and communicate with multiple cultural groups and be accepted within the groups, while maintaining comfortability as their own identified self
- **multicultural autonomous agricultural educator:** An educator who has the ability to connect two diverse cultural groups to the curriculum and find methods that allow the two cultural groups to collaborate and respect the differences that each brings to the collaboration
- **personal identifiers:** Distinguished characteristics for which the individual personally recognizes who they are and whom they associate with based on commonalities

- **social action approach:** The final stage of multicultural teaching reform where students gain curriculum knowledge while engaged/empowered to transform their gained knowledge to an action that helps their community regarding a social issue
- **social identifiers:** A traits or characteristics that an individual possesses, which are the basis for how others categorize themselves within social contexts and groups
- **transformation approach:** A structural change in the curriculum so that students are enabled to understand information from the perspective of diverse groups

Commonly Used Resources

- List of *Agricultural Education Magazine* issues relevant to diversity and inclusion:
    - Boone, H. N. (Ed.). (2014). Preparing the next generation of leaders. *The Agricultural Education Magazine*. Retrieved from https://www.naae.org/profdevelopment/magazine/archive_issues/Volume87/2014_09-10.pdf.
    - Boone, H. N. (Ed.). (2012). Serving students in agricultural education with special needs. *The Agricultural Education Magazine*. Retrieved from https://www.naae.org/profdevelopment/magazine/archive_issues/Volume84/2012_05-06.pdf.
    - Cano, J. (Ed.). (2012). Enhancing diversity. *The Agricultural Education Magazine*. Retrieved from https://www.naae.org/profdevelopment/magazine/archive_issues/Volume79/v79i1.pdf.
    - Ewing, J. C. (Ed.). (2018). Diversity and inclusion in agricultural education programs. *The Agricultural Education Magazine*. Retrieved from https://www.naae.org/profdevelopment/magazine/archive_issues/Volume90/2018%2005%20-%20May%20June.pdf.
    - Parham, E., & Vincent, S. K. (2018). Now is the time: effective strategies to connecting with African American students. *The Agricultural Education Magazine*. Retrieved from https://www.naae.org/profdevelopment/magazine/archive_issues/Volume90/2018%2005%20-%20May%20June.pdf.
    - Vincent, S. K. (2018). Let's create Cinderella stories. *The Agricultural Education Magazine*. Retrieved from https://www.naae.org/profdevelopment/magazine/archive_issues/Volume90/2018%2005%20-%20May%20June.pdf.
    - Vincent, S. K., Harper, T., & Tyler, Q. (2018). Are we truly serving all? keys to making a positive difference in the lives of all students. *The Agricultural Education Magazine*. Retrieved from https://www.naae.org/profdevelopment/magazine/archive_issues/Volume88/2016%2005%20-%20May%20June.pdf

- *Parent-Teacher Conferences: Strategies for Principals*, Teachers, and Parents Global Family Research Project, 2019
- Livermore, D. A. (2013). *Expand your borders: Discover ten cultural clusters.* East Lansing, MI: Cultural Intelligence Center, LLC

Figure Descriptions

Figure 5.1: Self identity includes awareness, attitude, and skill. Cultural discovery includes awareness, attitude, and skill. Together, self identity and cultural discovery make multicultural autonomy.

Figure 5.2: Level 1: The contribution approach. Level 2: The additive approach. Level 3: The transformation approach. Level 4: The social action approach.

Figure References

Figure 5.1: Multicultural autonomy growth model. Kindred Grey. 2023. CC BY 4.0.

Figure 5.2: Approaches to multicultural curriculum reform. Kindred Grey. 2023. Adapted under fair use from James A. Banks, "Approaches to Multicultural Curriculum Reform," 2008. https://scholarworks.umb.edu/trotter_review/vol3/iss3/5

References

Banks, J. A. (1993). Approaches to multicultural curriculum reform. In J. A. Banks & C. A. McGee Banks (Eds.), *Multicultural education: Issues and perspectives* (2nd ed.), 195–214.

Banaji, M. & Greenwald, A. (2013). *Blindspot: Hidden biases of good people*. Delacorte Press.

Cube, I. (1993). Check Yo Self [Song]. On *The Predator* [Album]. Priority Records.

Gay, G. (2003). *Becoming multicultural educators: Personal journey toward professional agency*. Jossey-Bass.

Gay, G., & Kirkland, K. (2003). Developing cultural critical consciousness and self-reflection in preservice teacher education. *Theory Into Practice*, *42*(3), 181–187. https://doi.org/10.1207/s15430421tip4203_3

Harris, R. P. (2001). Hidden voices: Linking research, practice, and policy to the everyday realities of rural people. *Journal of Rural Social Sciences*, *17*(1), 1–11.

Lortie, D. (1975). *Schoolteacher: A sociological study*. University of Chicago Press.

Schmitt, H. E. (1970). A model for preparing secondary teachers of agriculture for minority populations.

Journal of the American Association of Teacher Educators in Agriculture, 12(2), 20–29. http://doi.org/10.5032/jaatea.1971.02020

Vincent, S. K., & Drape, T. A. (2019). Evaluating micro expressions among undergraduate students during a class intervention exercise. *Nacta Journal, 63*(2), 133–139.

Vincent, S., & Westfall-Rudd, D. (2022, May 16–19). A Philosophical Approach to Obtaining Multicultural Autonomy as an Agricultural Educator. *American Association for Agricultural Education National Conference proceedings*, Oklahoma City, OK. https://aaea.wildapricot.org/resources/Documents/National/2022Meeting/2022AAAEPaperProceedings.pdf

6. DYNAMICS OF TEACHING

Tobin Redwine

Setting the Stage

On a hot and humid day in northern Uganda, I prepared my lesson plans and thought about the content I would soon be delivering to a group of agriculture teachers. As I sat in a modest classroom envisioning how the day's training would go, it began to rain. Normally, rain would not even register on my radar as an element to consider for effective teaching, but that day it was an absolute disruptor of my planned activities. As the rain beat down on the tin roof above my head, each drop clanged and slapped, echoing against the stone walls of the room. The increasingly loud drum beat pelted the room with a wall of sound. There was no way I could attempt to talk over the noise of the rain, so I immediately began to consider alternative locations for the day's activities.

I surveyed the area and available rooms, and I found that all the locations had the same challenge; so much noise would be a severe distraction for my teaching that day. Then, one of the participating teachers suggested moving to an outside area, underneath a canvas covering. The canvas muted the sounds of falling rain but presented other challenges. With no walls, nor electricity, I would need to reconsider my methods and approaches. I would have to change my strategy due to the environment, for my learners' attention and for my own abilities. Thankfully, we were able to pivot toward a new plan, including more hands-on interaction and discussion that engaged our learners, and met our objectives for the day. Now, every time it rains, I remember the noise in that classroom and the joyful session I had with teachers as we deeply discussed instructional strategies under a canvas, backed by the soft and gentle rain falling around us. I remember the effectiveness of change in a learning strategy and system.

Change is an essential and necessary element of effective teaching. Each classroom, each topic, each lesson, and each learner presents different characteristics that influence an instructor's approach. And just as each of these elements are fluid, effective instructors must also be fluid.

Objectives

The art of implementing change (whether planned proactively or adopted reactively) while teaching is called **dynamics of teaching**. In this chapter, you will explore a conceptual model designed to identify determinants (elements, or causal factors) that influence effective teaching and enhance your classroom dynamics. After completing this chapter, you will be able to meet the following learning objectives:

- Explain determinants that influence classroom environments, learner behaviors, and instructor qualities.
- Explain influence of relationships in establishing a dynamic classroom setting.

Introduction

To begin exploring the role of dynamics in teaching, consider the following Oxford Languages (n.d.) definition:

> ***Dynamics****: the forces or properties which stimulate growth, development, or change within a system or process*

To fully understand effective teaching, teachers must recognize the forces that are at play in their learning environments and respond to them in appropriate and meaningful ways. What features and facets in your classroom, your school, your system, and yourself influence the strategies you deploy? Do those factors influence what your learners find interesting? To attempt to answer how features in a system make learning interesting, we can look to an analogy provided by celebrated American author Kurt Vonnegut. He viewed stories as a tool to teach and offered insight into what makes a story interesting: their shape (Comberg, 2010).

Vonnegut described stories as having a shape. Imagine a graph. Picture an *x* axis that displays time from beginning to end of a story. Envision a *y* axis that portrays a continuum of good fortune at the top to bad fortune at the bottom. Each character's position on the graph changes—some move from good fortune to bad, some from bad to worse, some from good to better—as the story propels toward its end. These changes constitute a shape in a story. A character with a successful business or relationship at the beginning of a story might make a mistake and see their success turn to turmoil (causing a falling line on the graph), only to solve a problem and have their position on the continuum climb back toward the top. Vonnegut called that shape "man in a hole." Some stories involve an underdog, down on their luck, beginning their graph at the low end of the *y* axis, who climbs toward a happy ending as the story unfolds. That shape might be called "uphill climb."

Vonnegut's point was that stories become interesting *when their shape changes*. A straight horizontal line would imply a story with no challenge, no success, no change. That's an awfully boring story. Stories become interesting, celebrated, or beloved when their characters create dynamic shapes. Just as good stories need dynamics to be interesting, *good teachers need dynamics to be effective*. Teachers face a complex system of variables that influence successes and challenges in a classroom; this requires a dynamic, changing solution.

Overview of Theory and Practice

We have long known that changing one part of a system can cause change in another part of a system. Stanford psychologist Albert Bandura famously sought to describe whether it was "nature" or "nurture" that dictated a person's potential and performance. Instead of finding one or the other, his work led to the creation of a model that identified three **determinants**, or variables that can be changed to elicit a response. As we know, dynamics of teaching rely on understanding how changes affect teaching strategies. Bandura's model shows how these changes create cause-and-effect relationships. (See figure 6.1.)

First called Social Learning Theory and now known as Social Cognitive Theory, Bandura's model suggests that learning happens as a result of dynamic interaction of three determinants in a social exchange: environment, personal factors, and behavior (Bandura, 1989). Each determinant *influences* the other two determinants, and each determinant *is influenced by* the other two determinants. Introducing a change in one determinant will lead to changes in the others. Therefore, Social Cognitive Theory explains how dynamics in teaching happen; it happens by introducing change and expecting an effect in another part of the system.

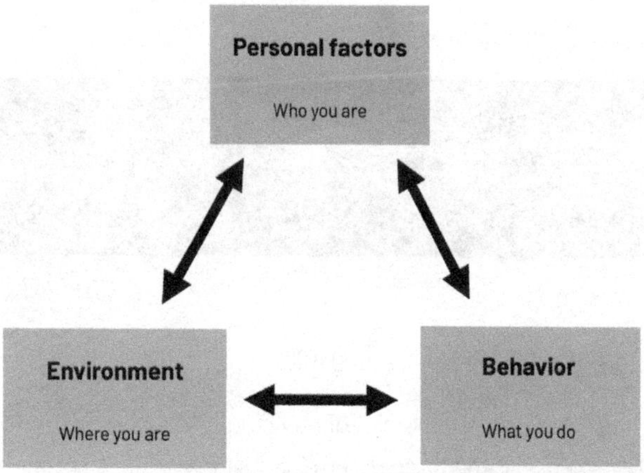

Figure 6.1: Determinants in the Social Cognitive Theory model.

Let's unpack each determinant of Bandura's model in a teaching context. In an education setting, consider the environment. The location, the physical space, the learning materials, and the orientation to peer learners all play key roles in characterizing the environment. The comfort (or lack thereof) of chairs upon which learners sit, the geography of the locations, the season of the year, the temperature of the space, the decorations, the navigability of the hallways, and countless other elements constitute the educational environment.

Personal factors describe preferences, personality traits, cognitive processes, and other intangible elements that make people who we are. Our values, expectations, identity, priorities, thoughts, emotions, beliefs, and knowledge base, among a myriad of other elements, constitute this determinant.

Finally, behaviors play an integral role in Bandura's model—both instructor and student behaviors. Actions, protocols, and other steps individuals take constitute our behaviors. The things we do and say define this determinant.

In summary, each determinant relies on and influences the other determinants.

- *Where you are* influences **what you do** and who you are.
- Who you are influences *where you are* and **what you do**.
- **What you do** influences who you are and *where you are*.

Social Cognitive Theory provides an excellent lens in examining the dynamics of teaching. The cause-and-effect model implies that *change leads to change*, and dynamics are all about change. If instructors want to see a change in our behavior, changing our environment will support or inhibit changed behavior. If instructors want to change students' perceptions, values, or personal factors, our behaviors must be crucially examined.

Learning Confirmation

Respond to the following items:

1. Define dynamics of teaching.
2. List the three determinants in Bandura's Social Cognitive Theory.
3. Characterize features in your own future teaching for each determinant.

Alternate assessment options:

1. Observe an instructor during a learning session. Describe elements of each determinant during the session (environment, instructor and learner personal factors, instructor and learner behaviors).
2. Prepare a learning session to deliver to peers that specifically introduces a change in behavior or environment as a tool for dynamics of teaching.

Applying the Content

After reading about the three-way influence of determinants, and how dynamics of teaching enhance educational activities, we are ready to put it all together. Make your own Bandurian map. Replicate the model from figure 6.1. Then, complete a personal inventory to show how determinants fit together. The suggestions below might inspire your application and reflection.

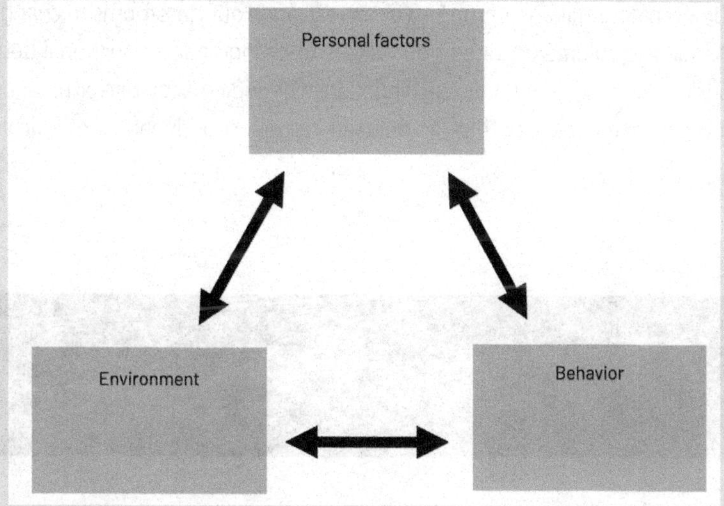

Figure 6.2: Social cognitive theory model (blank).

Environment

Describe an ideal learning environment in which you would like to teach. What are the physical characteristics? What are the spaces available and features? Compare that against an actual learning environment you have observed or learned in. Pay attention to elements in the environment that were helpful or were distracting as they each influenced the other determinants.

Behavior

Divide your behavior section of your map into two categories: learner and instructor. What behaviors do you expect from your learners? What behaviors do you expect of yourself? What behaviors are likely to occur? How can you measure or observe these behaviors to support assessment?

Personal Factors

List your strengths. If you aren't sure, begin by listing adjectives that you use to describe yourself. Ask a peer or someone you trust for additional adjectives. Then, list your interests and hobbies. Each of these contribute to a toolkit of strengths that you likely embrace as an instructor. Then take it further. What are your core values? What inspires you? What experiences or principles drive you the most?

Dynamic Strategies

Now, take a few moments to reflect on the map you created. How do items in each determinant influence each other? Based on what you know about the determinants in your context, generate a list of ideas that you can use to introduce change in each determinant. Perhaps we can rearrange our classrooms to change our environments, or maybe we can change the volume or tone of our voices to change our instructional behavior. The ideas we create here will be our toolkit for dynamics of teaching. Consider adding activities or breakout groups, games or exercises, discussion topics, and questions that can be used in support of dynamics of teaching.

Reflective Questions

1. Consider an instructor you believe to be an effective educator. What dynamics of teaching do they do well? Which of their strategies would you use in your own teaching and why? What strategies would be a challenge for you and why?
2. From your personal experience, list three potential pitfalls in a learning session. Consider ways to change a determinant to address each challenge.
3. Now that you have explained dynamics of teaching and determinants, what questions have we generated? Prepare a list of lingering questions. Use your list of questions as a catalyst for reaching out to a mentor, educator, or peer to engage in meaningful conversations and to pursue further resources for ongoing professional development.

Glossary of Terms

- **dynamics**: The forces or properties which stimulate growth, development, or change within a system or process
- **determinants**: Variables that can be changed to elicit a causal response in other variables

Figure Descriptions

Figure 6.1: Triangle diagram. Point 1: personal factors, meaning who you are. Point 2: behavior, meaning what you do. Point 3: environment, meaning where you are. Between each point of the triangle is a bidirectional arrow.

Figure References

Figure 6.1: Determinants in the social cognitive theory model. Kindred Grey. 2023. Adapted under fair use from Albert Bandura, "Human agency in social cognitive theory," 1989. American Psychologist, 44(9), 1175–1184. https://doi.org/10.1037/0003-066x.44.9.1175

Figure 6.2: Social cognitive theory model (blank). Kindred Grey. 2023. Adapted under fair use from Albert Bandura, "Human agency in social cognitive theory," 1989. American Psychologist, 44(9), 1175–1184. https://doi.org/10.1037/0003-066x.44.9.1175

References

Bandura, A. (1989). Human agency in social cognitive theory. *American psychologist*, 44(9), 1175.

Comberg, D. (2010, October 30). *Kurt Vonnegut on The Shape of Stories* [Video]. YouTube. https://www.youtube.com/watch?v=oP3c1h8v2ZQ

Oxford Languages. (n.d.). Dynamics. In *Lexico Powered by Oxford Languages*. https://www.lexico.com/en/definition/dynamics

7. PLANNING FOR EFFECTIVE INSTRUCTION

Amber H. Rice and Matt Mars

Setting the Stage

Robin looked around the classroom. They had just received their keys to the new agriculture program and they were overwhelmed by the piles of papers, textbooks, and equipment. The last agriculture teacher had taught in the school for over twenty years and was beloved by the entire district. Robin felt excited to embark on this new professional journey mixed with pressure to live up to the expectations of the students, school, parents, and community. As they opened the computer, Robin found no lessons plans, curriculum maps, or reference materials from the previous teacher. They thought,

> *I am not sure I am prepared for this. I have to teach a course on animal science, plant science, and agricultural engineering. I have some content knowledge in animal science, but I don't know enough about plant science or agricultural engineering to create lessons from scratch. How can I create curriculum for three courses with the school year starting in just three weeks? I don't even know where to start!*

Robin's excitement for their new position quickly devolved into panic as they viewed the first blank document to appear on their screen. They became overwhelmed and disheartened by the enormity of the task at hand: creating curriculum for their students.

Objectives

By the end of this chapter, learners will be able to:

- Explain the purpose of curriculum planning.
- Discuss the role of sociological and pedagogical theory on instructional design.
- Develop measurable learning objectives to guide instruction.
- Design curriculum maps for agriculture courses.
- Create comprehensive lesson plans.

Introduction

As experienced by Robin, knowing where to start when planning for effective instruction can sometimes be overwhelming. There are many factors to consider when crafting curriculum to meet the unique needs of your program and the students it serves. As you read through the chapter, we encourage you to begin with the end in mind. What skills, understanding, abilities, and behaviors do we intend for your students to gain at the conclusion of your class session, course, and agriculture program? How can we create curriculum that will lead to these outcomes and serve as the roadmap for highly effective instruction? What sociocultural and pedagogical factors influence how we go about the planning process? This chapter will address these important questions, and in doing so, provide a working guide for instructional planning.

What is Curriculum?

Curriculum planning is a vital and potentially daunting task. It first involves carefully thinking through what we need (and want) to teach as well as the best strategies for implementation. Second, it involves creating meaningful assessments that demonstrate student mastery of the content. Planning lessons is a critical component to being an effective educator. These instructional decisions inform and lead to what we call our curriculum.

Curriculum is defined as the sum of educational experiences in a course or series of courses (Merriam Webster, n.d.). It involves both the three-thousand-foot view of planning (i.e., curriculum maps) and the day-to-day planning (i.e., lesson plans). Curriculum encompasses all the courses we teach, how these courses build upon one another, and the content competencies we intend for students to learn. But before we can delve into the details of curriculum planning, it is important to consider the many factors, both internal and external, that can influence the decisions we make as educators when designing curriculum for our students.

Primary factors that influence curriculum:

- Disciplinary culture and field-based assumptions
- State and national education standards
- College and career readiness skills
- Employment opportunities and trends
- School and community context
- Facilities and resources
- Learners' age, prior knowledge, and background
- Scope of the course and program

Many factors influence curricular decisions. To begin, we must carefully consider the disciplinary culture of our subject area and field-based assumptions. Disciplinary culture can greatly impact not only the content we teach, but the lens through which we present that content. The section "Theoretical Considerations: From the Sociological to the Pedagogical," on theory related to planning, will dive more deeply into exploring the culture of our discipline.

Other relevant factors that influence curriculum include our state and national education standards. Each state sets standards for content differently. These standards are developed and disseminated at the state level and often include an outcome assessment to ensure students in every program are developing the identified competencies. In agricultural education for example, these state education standards may include technical agriculture content and professional skills or employability standards, and there are often have direct connections within the standards to leadership skills (FFA) and work-based learning outcomes (Supervised Agricultural Experience) (see chapters 10 and 11 for more information). In addition to state education standards, there are national education standards as well. National education standards are often adapted for use at the state level. These national standards are an excellent starting place if your state's standards provide less detail and direction than other states. As an example, the National Agriculture, Food, and Natural Resources (AFNR) Content Standards are divided into career pathways that represent groups of skills, knowledge, and interests related to an area of industry (The National Council for Agricultural Education, 2015). In agricultural education, career pathways can include agribusiness systems; animal systems; biotechnology systems; environmental service systems; food products and processing systems; natural resources systems; plant systems; and power, structural, and technical systems (The National Council for Agricultural Education, 2015). These career pathways are most often served through a series of courses at the secondary level.

In addition to state and national education standards, there is an emphasis in many disciplines on college and career readiness skills. These skills are embedded throughout curricula with the intent of preparing all students to graduate from high school ready for college, careers, and life. Emphasis on these skills should be included throughout the entire curriculum. In addition, your state may have specific metrics for student attainment of these essential skills. For example, in agricultural education, college and career readiness skill attainment includes employment opportunities that match trends within the agricultural industry from a national, state, and local community perspective.

As educators, we also need to carefully consider our community context. Curricular decisions are influenced by the specific community in which an education program is located. It is important to develop an awareness of the unique characteristics in our communities, such as employment, history, and the influence of relevant stakeholder groups. Creating a curriculum group is one way to seek input from the community, industry professionals, leaders, and stakeholders unique to your specific location. We also need to consider our school context. School context includes the type of school in which we are employed (e.g., public, private, or charter school) and if the school is located in an urban, suburban, or rural environment. Each of these characteristics plays a role in our curricular decision-making.

We must also take into account the facilities, equipment, and other resources that are at our disposal for influencing student learning. While additional resources can sometimes be acquired, taking a careful inventory of the available items can aid in making purposeful curricular decisions. For example, if your school has a greenhouse, you may wish to include science content related to seed germination as a means of encouraging active learning. The facilities and equipment need not dictate the content we choose to teach but can serve as important factors in how we deliver content to our students in meaningful ways.

We must also consider the learners themselves. Learners' ages, backgrounds, needs, interests, and prior knowledge are critical to curriculum development since our students are our audience! The backgrounds and abilities of our learners can guide how quickly we move from one content topic to another, the depth of knowledge we desire to teach, and the teaching methods we use to properly align with the developmental levels of our students. Additionally, we must carefully consider the needs and interests of our learners to ensure our curriculum is adequately preparing them for their future.

Last, it is imperative that we recognize and remain attentive to who we are as educators and the perspectives and potential biases our own backgrounds inherently bring to the curriculum planning process. The role of the educator is to expand the knowledge levels and enhance the capacities of learners to prepare them to critically think for themselves. Therefore, it is our responsibility to take appropriate measures in minimizing the influence of our own values and agendas throughout the planning and content delivery process.

Each of these factors plays an essential role in curricular decision-making. To be effective, educators need to continually update their knowledge about each of these factors as they change over time, keeping pace with innovation, cultural, and economic shifts in our profession and in our communities. Keep each of these influential factors in mind as we further consider with greater specificity the development of curriculum maps, unit plans, and lesson plans for instruction.

Planning provides a thoughtful blueprint to guide our instruction. However, remember the learning environment is dynamic and ever-changing. Sometimes the best laid plans must be adapted to meet the unique and fluid needs of individual learners, specific learning environments, and distinct communities in real-time. Take planning seriously and with a willingness to deviate from the *script* when needed.

Background: Theory and Research

Theoretical Considerations: From the Sociological to the Pedagogical

We all bring a range of academic and personal perspectives and biases to our professional practices. We were socialized to view disciplinary content through field-based lenses that are specific to the programs of study that prepared us to be educators. Every discipline is anchored in its own culture with consequent socialization being enacted through established moral orders (Harré, 1983). Moral orders dictate what is valued and considered to be on one hand acceptable and desirable and on the other hand objectionable and improper. In the academic context, the moral orders that underpin disciplinary cultures heavily influence how we as practitioners come to view what is credible knowledge, what worthwhile learning looks like, and what desirable outcomes are in terms of how students make sense of and apply what is being taught (Shotter, 1994). These orders are so deeply embedded within disciplines that they are passed on and enacted as basic assumptions (Schein, 2010)—cultural elements that are taken for granted and, over time, practiced with little to no awareness. They become the "it's just what we do" of practice. The insidiousness of moral orders confines the imagination of practitioners to disciplinary-based boundaries of what is acceptable practice and legitimate positions and viewpoints (Ylijoki, 2000).

Conflicts and disconnects between the disciplinary cultures and moral orders of educators and the worldviews of learners can compromise productive and meaningful learning. Sometimes obvious and other times not-so-obvious barriers to learning can arise as learners are forced to reconcile conflicts between their worldviews and those of their families with those of their educators. The results can be a breakdown in their learning or complete disengagement from the course and its content.

One approach to proactively minimizing such unintended fallout of moral order bias is to embed routine reflection and self-discovery in the curriculum planning process. As educators, we often promote the importance and value of fostering critical thinking among our learners. We, too, should be actively engaged in critical thinking as a matter of routine practice. Lai (2011) identified the fundamental elements of critical thinking as "attitudes or habits of mind, [that] include open and fair-mindedness, inquisitiveness, flexibility, a propensity to seek reason, a desire to be well informed, and a respect for and willingness to entertain diverse viewpoints" (p. 2). Consistent with this understanding, educators must identify the influence of their own biases in their curriculum and, through doing so, include strategies for creating constructive spaces in which students are able to make sense of content relevant to their own worldviews. Such strategies might include reflective journaling, debate exercises, and diversified readings that make it clear there is rarely one correct way of seeing and understanding the phenomena being learned. Management and industry consultants often say *culture eats strategy*. Indeed, curriculum planning inherently involves strategy. Without considering the influence of the cultures that shape our worldviews as educators, the strategies we build into our planning will be incomplete and detrimental to the openness and inclusivity of our learning environments.

Complementary to the sociological considerations, we must also consider pedagogical theory and how it applies to our curriculum design. Our own beliefs as educators heavily influence and shape the content we choose to develop and how we choose to deliver that content (Rice & Kitchel, 2017a). Our integrated beliefs systems that shape our pedagogical content knowledge (i.e., knowledge for teaching) include our beliefs about the purpose of our discipline, our beliefs about specific content areas, and our general beliefs about teaching and learning. These beliefs interact and inform one another, creating our personal integrated beliefs system (Rice & Kitchel, 2017a, 2018).

Educator beliefs about the purpose of our discipline can include career preparation, college preparation, literacy, life skills, and learner individualization, which encompasses the other four belief categories (Rice & Kitchel, 2017b). Educators may hold multiple beliefs about the purpose of their discipline simultaneously. These beliefs impact a variety of instructional decisions that influence learning outcomes. For example, if an educator holds the belief that the purpose of their discipline is career preparation, they may incorporate more manual skill learning outcomes into the curriculum and may emphasize hands-on learning applications. Conversely, if an educator holds the belief that the purpose of their discipline is literacy development, they may incorporate more discussions and assignments that focus on developing knowledge for informing consumers.

Educator beliefs about specific content areas also shapes our planning and practice (Rice & Kitchel, 2017a). Within plant science education, for example, educators may believe the purpose of a school greenhouse is to serve as a laboratory for plant science experiments and may treat the greenhouse as an extension of the classroom to aid in developing science knowledge and skills. Conversely, educators may believe the purpose of the school greenhouse is a production facility, and consequently, they may focus instruction on the viability of the plants produced in the greenhouse and may engage in lessons related to marketing and sales.

Finally, educator beliefs about teaching and learning beyond the scope of a specific disciplinary context also influences our curriculum development (Rice & Kitchel, 2017a). Educator philosophical beliefs about teaching and learning in our discipline include believing that it is the educator's responsibility to be a lifelong learner, believing in the importance of educator reflection on the learning process, believing that students play a role in determining the content to be delivered, and believing that students learn best through real-life applications of concepts (Rice & Kitchel, 2017a). As educators, we must engage in meaningful reflection about our beliefs as they have the power to shape curriculum planning and instruction at the micro and macro levels.

Cognitive Levels and Bloom's Taxonomy

Bloom's Taxonomy is a common educational framework that guides curriculum development through the creation of measurable learning objectives relevant to hierarchical levels of thinking (Anderson & Krathwohl, 2001; Bloom & Krathwohl, 1956). Bloom's Taxonomy focuses on the cognitive or knowledge-based learning domain and contains the following levels of cognition, beginning with the least complex and scaffolding upward: remember, understand, apply, analyze, evaluate, and create (Anderson & Krathwohl, 2001). Each level can be associated with various action verbs to guide measurable development of learning objectives and the subsequent pedagogical approaches and techniques. In addition to the cognitive domain, education theorists have also outlined levels for the affective or emotion-based learning domain and the psychomotor or action-based learning domain (Bloom & Krathwohl, 1956; Harrow, 1972; Krathwohl et al., 1964). All three domains of learning (cognitive, affective, and psychomotor) are equally important in curricular development. We will focus on the cognitive domain as we delve into writing learning objectives.

Writing Measurable Learning Objectives

The first step in curriculum planning is crafting measurable learning objectives to guide instruction. These learning objectives provide direction for the educational process and the standards by which learner performance (i.e., competence) can be assessed (i.e., evaluated). Heinrich et al.'s (1996) model for writing learning objectives provides a concrete process for creating each component (A, B, C, D) of an objective.

"A" is the *audience* in the ABCD learning objective model. The "A" component answers the question "Who is the learner?" All learning objectives are learner-centered as your desired outcome is the for the learner to be able to accomplish the objective. Whether explicitly written in the learning objective or not, all our objectives should begin with "The learner will be able to ..." In addition to guiding your curriculum planning as the educator, objectives should be shared with students as they are your audience.

"B" is the *behavior* in the ABCD learning objective model. The "B" component answers the question "What should the learner be able to do?" All learning objectives provide a clear and concise statement of the performance required of the learner. This aligns with the competency or expected outcome of the lesson. The behavior is written as a measurable action verb. Example action verbs include describe, identify, explain, differentiate, synthesize, calculate, and create. These verbs are chosen to align with the cognitive levels of Bloom's Taxonomy.

"C" is the *condition* in the ABCD learning objective model. The "C" component answers the question "Under what conditions do you want the learner to be able to do it?" The condition contains the settings or circumstances within which the learner is to perform or demonstrate the behavior. Example conditions include information that the learners may be provided that will direct the action; the environment in which the performance must be demonstrated; or the equipment, supplies, or materials which the learner is provided. The chosen conditions directly align with our teaching methods used to deliver the content.

"D" is the *degree* in the ABCD learning objective model. The "D" component answers the question "How well must it be done?" The degree indicates the basis upon which the performance will be evaluated and is used to assess how well the learner must perform to be judged as competent. Example degrees include percent accuracy, maximum number of errors, standards for excellence, reference to other materials (e.g., rubrics) which identify specific criteria, or a combination of multiple factors.

Some example objectives that contain all four components of the ABCD objective model for a lesson on soil texture include:

- Following the lecture, learners will be able to identify the three sizes of soil particles, with 100 percent accuracy.
- Following the discussion, learners will be able to describe the importance of soil texture, including all four factors.
- Given the soil texture triangle, learners will be able to calculate the soil type, with 90 percent accuracy.
- Following the demonstration, learners will be able to demonstrate the soil ribbon test, including all components on the teacher provided checklist.

Clear, concise, and measurable learning objectives are the cornerstone for efficient and effective planning. Think of learning objectives as the road map for curriculum design.

Mapping Curriculum: Overview

When mapping curriculum, the order in which content flows from one topic to the next is important. We must think about our curriculum from a variety of perspectives including across our total program (i.e., multiple courses) and across a single course (i.e., units of instruction). If you are employed in a single teacher program, this may be largely an individual effort at the program level. If you are employed in a multi-teacher program, you must collaborate with other educators in your program to ensure the curriculum flows from course to course. Finally, if you are engaging heavily with other content areas within your learning environment (e.g., biology or mathematics) you will be collaborating with those educators when designing curriculum.

From the program perspective, it is important that your courses build upon one another. For example, we may have an introductory level course geared toward novice learners that introduces a variety of content. From there, we may offer a course that is geared toward intermediate learners, or we may have various specialty courses open to students across learning levels. The course scope and sequence are school, program, and learning environment dependent. For example, you may consider the career pathways for AFNR (National Council for Agricultural Education, 2015) when creating courses and paying attention to the content topics that are included in each specific course. We may decide to introduce some content topics in one course and build upon those topics in subsequent courses. Depending on the size of our programs and the number of educators collaborating, we may offer a variety of courses in which learners can engage. Keep in mind, each of the highlighted factors are influential when making programmatic curricular decisions.

Mapping Curriculum: Across a Single Course

For each course we teach, we will create a curriculum map that outlines the following components of the course: individual units of instruction, individual lessons for each unit, time frame for each unit, and time frame for each lesson and topic. Our curriculum maps are our guide to instruction for that specific course. We create a curriculum map before designing individual units of instruction or lessons plans.

Units of instruction are the building blocks for our curriculum map. These units encompass various content topics that hang together and will lead us to individual lesson plan creation. Using content standards as a guide, consider the broad areas of instruction you would introduce within an individual course to meet course goals. For example, in an animal science course, we may have units of instruction that include, but are not limited to, introduction to animal science, animal science careers, animal breeds, animal anatomy, animal nutrition, animal husbandry and behavior, animal health and diseases, and animal reproduction.

After deciding the broad units of instruction to include in the course, we will outline the time frame for each unit. Consult supervisory personnel for length of time expected per course when making curriculum decisions. Units of instruction can span various lengths of time, but two to six weeks in length is common. The length of a unit depends on the overall time we have to teach within the course as a whole, how in-depth we plan to go with content topics, the prior knowledge of our learners, and the learning objectives we must accomplish within that unit.

For each unit of instruction, create a unit outline illustrating the individual lessons that encompass the unit. A single unit will consist of a varying number of individual lessons. For example, within an animal anatomy unit, we may include lessons on external anatomy, circulatory system, endocrine system, nervous system, and skeletal system. For each of these lessons we will create learning objectives to guide the lesson plan creation.

In addition to the previous considerations, we must also account for additional time factors within our instruction. We may have ongoing activities that span various units of instruction, time allotted to utilize facilities like a land lab or greenhouse, field trips planned within the course, large projects with authentic assessments, or time for traditional assessments. It is easy to overestimate the instructional time needed if we do not take into consideration the time it requires to incorporate these activities. We should also consult necessary calendars for dates, such as holidays or other events, that can impact time for instruction.

Curriculum Map Template

This template will serve as a guide for creating curriculum maps for each of our courses (see table 7.1). Adapt this template as needed to address the unique aspects of our learning environments.

Week	Unit	Lessons (Instructional Periods)
1		
2		
3		
4		
5		
6		
7		
8		
9		

Table 7.1: Curriculum map template.

Context and Time Length for Lessons

For each lesson plan created, we must consider the context of that lesson and the length of time needed for delivery. These curricular decisions are guided by our curriculum map, unit outline, and the learning objectives we have outlined for each lesson. Some lessons will be fully taught within a single session, while others may last multiple sessions. Within each lesson, consider how the content connects to previous and future lessons. Facilitate students making connections between content topics from lesson to lesson and from unit to unit to establish meaning and relevancy. Content should build within and between lessons, units, and courses.

Utilizing Lesson Planning Resources

Utilizing our resources effectively is vital when developing curriculum. It is likely that another educator has already developed a lesson for the content we are preparing. Take advantage of existing resources to avoid *recreating the wheel*. Lesson planning resources can include textbooks, human resources, digital resources, curriculum groups, national organizations, educators in various disciplines, and social media groups, to name a few. For example, in agricultural education, communities like the National Association of Agricultural Educators (NAAE) facilitate the sharing of lesson plans and activities between agriculture educators at the national level through communities of practice (NAAE, 2021).

There are also curriculum packages available in most disciplines. For example, in agricultural education, there are curriculum packages available within iCEV and Curriculum for Agricultural Science Education (CASE) (CASE, 2021; iCEV, 2021). Curriculum packages can be excellent resources when creating curriculum. For each curriculum resource we utilize, carefully consider the validity of the information presented. Do not assume the information is accurate and current just because it is made available or created by another educator. Adapt any resources utilized to meet the needs of your specific learning environments.

Creating the Lesson Plan

When creating lesson plans, it is important to include key components that will appear in every lesson we deliver. Use a lesson plan template to organize instruction (see table 7.2). In the following section, we share common components of a lesson plan template and provide an example template that can be adapted for individual use. Keep in mind, many learning environments may provide a unique required lesson plan template.

Lesson Plan Template

This lesson plan template is a guide to creating lesson plans. Adapt this template as needed to address the unique aspects of the learning environment.

Lesson Plan	
Educator:	
Course:	
Unit Title:	
Lesson Title:	
Audience:	
Estimated Time:	

Standards Connections:

1.
2.
3.

Enabling Objectives: *Learners will be able to...*

1. [List here using ABCD format]
2.
3.

Materials and References:

Tools, Supplies, and Equipment:
• [List]

Handouts/Digital Materials:
• [List]

Resources/References:
• [List]

Accommodations:

1.
2.
3.

Key Terms:

• [List]	• [Define]

Bell Work:	**Estimated Time:**
• • •	

Interest Approach:	**Estimated Time:**
Hook: Felt need to learn: Transition to content: Overview of lesson objectives:	

Objective 1:	**Estimated Time:**

Lesson Plan	
Instructor Directions / Materials / Teaching Procedure	**Brief Content Outline**
Method: Idea for Review: Formative Assessment:	
Objective 2:	**Estimated Time:**
Instructor Directions / Materials / Teaching Procedure	**Brief Content Outline**
Method: Idea for Review: Formative Assessment:	
Objective 3:	**Estimated Time:**
Instructor Directions / Materials / Teaching Procedure	**Brief Content Outline**
Method: Idea for Review: Formative Assessment:	
Summary / Conclusion	**Estimated Time:**
[Insert summary here]	
Assessments	
Formative (for each objective): • • •	
Summative: •	

Table 7.2: Lesson plan template.

Lesson Plan Components

The first component of the lesson plan template is the *background information*. This includes the educator's name, the course or session title, the unit of instruction title, the lesson plan title, the audience served, the number of learners served, and the estimated time allotted for the lesson.

The second component of the lesson template is the *standards connections*. Where applicable, connect the content of the lesson to any oversight standards. This may include a state's disciplinary-specific content standards, professional skills or employability standards, or other core content standards beyond the discipline (e.g., biology, mathematics, economics, social studies, and English). When adding standards connections to a lesson plan, be sure that the lesson is addressing the specific standards outlined. If at first it appears that no standards apply to the lesson content, we may need to rethink the content depending on our content oversight agency. Some lessons may connect to multiple content standards.

The third component of the lesson plan template is the *enabling objectives*. This is where we outline the objectives we have created for the lesson, using the ABCD format discussed previously in the chapter (see "Writing Measurable Objectives"). Enabling objectives will build upon one another, be written for the audience, and scaffold across various levels of Bloom's Taxonomy. Enabling objectives serve as the roadmap for our content delivery throughout the lesson.

The fourth component of the lesson plan template is the *materials and references*. This is where we identify all materials, tools, supplies, and equipment needed for the lesson. We will also list all handouts or digital materials needed for the lesson. Creating the list of materials will assist in organizing the needed supplies for successful delivery of the lesson. Use this list to prepare for the session by purchasing materials, making copies, and uploading digital resources. Within this section, identify all sources of information used to create the lesson. This includes, but is not limited to, websites, lesson plans from other sources, curriculum packages, and textbooks. By outlining sources of information in a reference list we can easily return back to the original source at a later date to ensure current and accurate information for content delivery.

The fifth component of the lesson plan template is the *key terms*. This is where we identify all vocabulary terms and their definitions as they appear in the lesson. Incorporating vocabulary within the lesson is an important consideration as we prepare learners to correctly use common terminology related to the content.

The sixth component of the lesson plan template is the *accommodations*. This is where we outline any accommodations that will be made to serve learners. This can include changes that will be made during the lesson for students with IEPs, 504s, as well as any needed accommodations for all learners. Completing this section while carefully considering the needs of your unique learners is a critical component of the lesson planning process.

The seventh component of the lesson plan template is the *bell work* (commonly referred to as a bellringer and various other names). This is where educators outline a short engagement activity for learners to complete prior to the formal start of the session. Bell work can be in the form of review questions, questions to elicit prior knowledge, quotes to respond to, or additional ideas. The purpose of bell work is to activate learners' minds immediately upon entering the learning environment. Providing learners with bell work also allows educators the necessary time to take attendance or attend to other required tasks. Bell work is typically five minutes or less of the total instructional time.

The eighth component of the lesson plan template is the *interest approach*. This is the formal start to the lesson. The purpose of an interest approach is to create in our learners a felt need to learn the content. A typical interest approach will hook the learner, establish motivation for learning the content, and transition learners' minds to the content, and it often ends with an overview of the enabling objectives. Interest approaches are typically five minutes in length but may vary depending on the lesson and the interest approach content. Designing interest approaches after our lesson content is planned allows for effective connections between the interest approach and the rest of the lesson.

The ninth component of the lesson plan template is the *content*. This is where we describe the outline of our lesson content that corresponds to each enabling objective. In the example template (see table 7.2), each content section is listed with a place to enter the enabling objective and an estimated time for completion. The template is divided into two columns. The left-hand column is for the educator to outline any directions, materials, and procedures. Educators should identify the teaching method being used to deliver the content and the formative assessment being used to check for understanding at the end of the content delivery for each objective (see the twelfth component of the lesson plan template related to assessment). The right-hand column is for the educator to outline the content delivered. Educators should include visual slides, questions to be asked of learners, and directions for learning activities. At the end of each content outline, we surface an idea to review the content taught for that objective. The content section of our lesson plan template will contribute to the bulk of the lesson plan. We encourage educators to carefully think through the content, method(s) used for delivery, visual aids, engaging questions and activities, assessments, and reviews of the content for each enabling objective.

The tenth component of the lesson plan template is the *application*. This is where educators design instructional activities for learners to apply the content that has been taught. This is not a stand-alone section on the lesson plan template, but instead should be incorporated into each content section corresponding to the enabling objectives. Students learn best when they have an opportunity to practice and apply content, so designing meaningful applications throughout our lessons encourages students to move past "remember" (in Bloom's Taxonomy) to reach higher cognitive levels. Facilities such as land labs or greenhouses may be utilized to purposefully apply content but are not required for quality instruction. For in-depth lab application integration, see the additional lesson plan considerations section on lab applications.

The eleventh component of the lesson plan template is the *summary/conclusion*. This is where we wrap-up the lesson and engage learners in meaningful reflection on the content. The summary/conclusion should review the most salient points of instruction and allow for learners to process the content. Discussions and short journal reflections are strategies we can use to engage learners in summarizing the content. Effective summary/conclusions tie directly back to the interest approach that was used to introduce the session.

The twelfth component of the lesson plan template is the *formative and summative assessments*. Formative assessments are designed to check for understanding of the content as the lesson unfolds. We include a formative assessment at the conclusion of each enabling objective. Formative assessments allow us as the educator to determine learners' comprehension of the content in real time, so we can make adjustments or reteach as necessary. Formative assessments include peer discussion, mini-reflection papers, learners' self-rating procedures, polling, discussions, and example problems, to name a few. Summative assessments are designed to measure comprehension at the end of the lesson or unit, across multiple enabling objectives. Summative assessments can include projects, papers, presentations, skill demonstrations, exams, and case studies, to name a few. Assessments must be chosen based on content being delivered, needs of learners, and oversight agency requirements. Assessments can be traditional, in the form of quizzes and exams, or authentic, in the form of projects, presentations, and applications of learning. Finally, we engage students in a summative assessment of an entire course through an engaging capstone-type item.

Additional Lesson Plan Considerations

- Integrating FFA and SAE (see chapters 10 and 11 for further information)
- Lab applications (see chapter 12 for further information)
- Universal design considerations

Specifically for agricultural education, it is important to consider connections to FFA (i.e., leadership) and SAE (i.e., experiential or work-based learning) directly within the lesson plan. FFA connections can include lessons specifically on FFA content (e.g., history, emblem, official dress, and the FFA Creed) and the incorporation of FFA components through career development events (CDEs) and leadership applications (e.g., conflict management, teamwork, and parliamentary procedure). See chapter 10 for more information on FFA and its role in our discipline.

SAE connections can include lessons specifically on SAE content (e.g., areas of SAE, proficiency awards, purposeful selection of SAE projects) and the incorporation of SAE components (e.g., record keeping, budgeting and finance, workplace employability skills, agricultural literacy, careers in the industry, and workplace safety) (see The National Council for Agricultural Education, 2017). See chapter 11 for more information on SAE and its role in our discipline. By incorporating FFA and SAE directly into lesson plan development we can maintain the intracurricular nature of the three-component model of agricultural education that sets it apart from other education disciplines.

When creating dynamic, application-driven curriculum, educators often incorporate laboratory instruction and technical skill development. To facilitate this within our lesson plans, we utilize a lab application guide as a supplemental handout to the lesson plan. The lab application guide contains step-by-step how-to instructions for performing the lab or skill. Lab application guides can be useful in the learning environment to provide learners with key aspects of the lab. The lab application guide will contain materials needed, steps to complete, key points to emphasize, safety considerations (if applicable), and illustrations to illuminate complex steps. We recommend surfacing the use of the lab application guide directly in the lesson plan content and then attaching the guide as a supplemental document to support the lesson. Lab application guides can support a variety of long-term projects, skills, or labs that require multiple steps to complete.

Within each of our lessons, educators need to incorporate universal design practices. This involves making thoughtful design choices to incorporate inclusive practices in our instruction to serve all learners. Universal design is simply the design of products, environments, and communication to be usable by all learners without the need for individualized adaptation (Center for Universal Design, 1997). Universal design benefits learners of all ages and abilities and can include both physical and nonphysical features of the lesson. An example of universal design includes using captions or subtitles on recorded videos that will be shown to learners. Captioning assists learners with hearing impairments, but can also support learners with language barriers, those who are learning new content and vocabulary, those who are learning in a room with distractions or competing noises, and those who are taking comprehensive notes. Adding captions or subtitles is a small change that can benefit every learner. Universal design saves time because educators are making fewer individual modifications to serve specific learners.

The seven key principles of universal design, as articulated in the professional literature, are equitable use, flexibility in use, simple and intuitive use, perceptible information, tolerance for error, low physical effort, and size and space for approach and use (Center for Universal Design, 1997). To attend to each of the principles we ask ourselves the following questions as we design curriculum.

- Did I design with all potential groups and motivations in mind?
- Does my design work for learners who may need to approach things differently than I would?
- Is my design easy to use regardless of socioeconomic status, culture, and experience?
- Did my design communicate necessary information effectively, in a way that considers sensory issues and the environment?
- Did I design in a way that minimizes or allows accidental learner error?
- Did I design in a way that advantages the able bodied?
- Is my design appropriate for the environment and learners' mobility?

By implementing universal design, educators can better serve learners through practices that promote accessibility and inclusion for all.

Learning Confirmation

Now that quality education products have been created, reflect on the following questions as the education products are reviewed.

Curriculum Map

1. Did I consider the various factors that impact curricular decision-making?
2. Did I reflect on my own perspectives and bias and how they influence the planning and delivery processes?
3. Did I consider the other courses in my program and how the content of this course will build on and connect to other courses?
4. Did I consult with other educators or content experts in my learning environment to ensure congruency?
5. Did I consider career pathways and state standards as applicable and how they will be achieved within my sessions?
6. Did I inventory the supplies and equipment available to aid instruction?
7. Did I outline the individual units of instruction and indicate the time frame for each unit?
8. Did I outline the lessons for each unit of instruction and the time frame for each lesson?
9. Did I consult any guiding schedules to take into account holidays, community events, and dates that influence my sessions?

Lesson Objectives

1. Did I follow the ABCD format outlined in the chapter?
2. Do each of my learning objectives include a behavior written as an action verb?
3. Do my learning objectives scaffold across various levels of Bloom's Taxonomy?
4. Did I consider the three domains of learning in objective creation?
5. Did I consult state standards and national career pathways as applicable?
6. Did I indicate the condition in which the learners will achieve the objectives?
7. Are each of the learning objectives measurable according to a specified degree?

Lesson Plan

1. Did I identify and utilize quality lesson planning resources?
2. Did I include the following components of a lesson plan: background information, standard connections, enabling objectives, materials and references, key terms, accommodations, bell work, interest approach, content, application, summary/conclusion, and formative and summative assessments?
3. Did I integrate FFA and SAE into my lesson plan creation when appropriate?
4. Did I consider the use of a lab application guide to supplement instruction when appropriate?
5. Did I incorporate the seven principles of universal design to serve all students?

Applying the Content

Using the knowledge we have gained throughout this chapter, let's apply the concepts through creating education products. Using the templates and chapter content as a guide, create the following:

1. Curriculum map (for a single course)
2. Lesson objectives (for a single lesson)
3. Lesson plan (for a single lesson)

Reflective Questions

1. What curriculum factors will we consider that influence our instructional design and practice?
2. What disciplinary-, field-, and experience-based perspectives and biases influence and threaten to bias our instructional practices?
3. What are our personal beliefs about the content we deliver?
4. How can we create congruency across courses, units, and lessons within our instruction?
5. How will we utilize objectives to create a roadmap for instruction?
6. What key components will we include in our lesson plan creation?
7. What resources can we explore further to assist with curriculum design?
8. How can we meaningfully integrate FFA and SAE into agricultural curriculum?
9. How can we apply principles of universal design to our learning environments, communication with learners, and teaching products?

Glossary of Terms

- **curriculum:** Sum of education experiences in a course or series of courses (Merriam Webster, n.d.)
- **Bloom's Taxonomy:** Common educational framework that guides curriculum development through the creation of measurable learning objectives relevant to hierarchical levels of thinking (Anderson & Krathwohl, 2001; Bloom & Krathwohl, 1956)
- **learning objectives:** Roadmap to provide direction for the educational process and the standards by which learner performance can be assessed
- **audience:** The "A" in the ABCD learning objective model that answers the question "Who is the learner" (Heinrich et al., 1996)
- **behavior:** The "B" in the ABCD learning objective model that answers the question "What should the learner be able to do?" (Heinrich et al., 1996)
- **condition:** The "C" in the ABCD learning objective model that answers the question "Under what conditions do you want the learner to be able to do it?" (Heinrich et al., 1996)
- **degree:** The "D" in the ABCD learning objective model that answers the question "How well must it be done?" (Heinrich et al., 1996)
- **curriculum map:** Document that outlines the following components of a course: individual units of instruction, individual lessons for each unit, and the time frame for each lesson and topic
- **units of instruction:** Building blocks of the curriculum map that encompasses various connected topics that hang together
- **unit outline:** Document that outlines the individual lessons that encompass a unit
- **lesson plan template:** Document that serves as guide to creating a lesson plan
- **Universal Design:** The design of products, environments, and communication to be usable by all learners without the need for individualized adaptation (Center for Universal Design, 1997)

References

Anderson, L. W., & Krathwohl, D. R. (Eds.)(2001). *A taxonomy for learning, teaching, and assessing: A revision of Bloom's taxonomy of educational objectives.* Longman.

Bloom, B. S., & Krathwohl, D. R. (1956) *Taxonomy of educational objectives: The classification of educational goals by a committee of college and university examiners.* Handbook I: Cognitive Domain. Longmans, Green.

Center for Universal Design. (1997). The principles of universal design. Retrieved February 25, 2021, from https://design.ncsu.edu/research/center-for-universal-design/

Curriculum for Agricultural Science Education (2021). Curriculum standards alignment. Retrieved March 22, 2021, from https://www.case4learning.org/curriculum/curriculum-standards-alignment/

Harré, R. (1983). *Personal Being*. Oxford: Basil Blackwell.

Harrow, A. (1972). *A taxonomy of psychomotor domain: A guide for developing behavioral objectives.* David McKay Co.

Heinrich, R., Molenda, M., Russell, J. D., & Smaldino, S. E. (1996). *Instructional media and technologies for learning*. Merrill.

iCEV (2021). Standards aligned curriculum agricultural science. Retrieved March 22, 2021, from https://www.icevonline.com/curriculum/agricultural-science

Krathwohl, D. R., Bloom, B. S., & Masia, B. B. (1964). *Taxonomy of educational objectives, the classification of educational goals*. Handbook II: Affective Domain. David McKay Co.

Lai, E. R. (2011). *Critical thinking: A literature review*. Pearson Education.

Merriam-Webster. (n.d.). Curriculum. In Merriam-Webster.com dictionary. Retrieved January 29, 2021, from https://www.merriam-webster.com/dictionary/curriculum

National Association of Agricultural Educators. (2021). Communities of practice. Retrieved March 22, 2021, from https://www.naae.org/

The National Council for Agricultural Education. (2017). National AFNR content standards. Retrieved January 29, 2021, from https://thecouncil.ffa.org/afnr/

The National Council for Agricultural Education. (2015). SAE for all. Retrieved January 29, 2021, from https://saeforall.org/educator-resources/

Rice, A. H., & Kitchel, T. (2018). Agriculture teachers' integrated belief systems and its influence on their pedagogical content knowledge. *Journal of Agricultural Education*, *59*(1), 51–69. http://doi.org/10.5032/jae.2018.01059

Rice, A. H., & Kitchel, T. (2017a). Shaping pedagogical content knowledge for experienced agriculture teachers in the plant science: A grounded theory. *Journal of Agricultural Education*, *58*(4), 50–64. http://doi.org/10.5032/jae.2017.04050.

Rice, A. H., & Kitchel, T. (2017b). Teachers' beliefs about the purpose of agricultural education and its influence on their pedagogical content knowledge. *Journal of Agricultural Education*, *58*(2), 198–213. http://dx.doi.org/10.5032/jae.2017.02198

Schein, E. H. (2010). *Organizational culture and leadership* (4th ed.). Jossey-Bass.

Shotter, J. (1994). *Conversational realities*. Sage Publications.

Ylijoki, O. H. (2000). Disciplinary cultures and the moral order of studying: A case-study of four Finnish. *Higher Education*, *39*, 339–62. https://doi.org/10.1023/A:1003920230873

8. DELIVERING CONTENT WITH TECHNOLOGY

OP McCubbins and Annie Specht

Setting the Stage

It's mid-March, and your students are preparing for spring break. As you send them out the door on Friday, backpacks laden with homework and textbooks, the local news is abuzz with stories about a deadly virus currently spreading throughout the United States. Within twenty-four hours, the superintendent has announced that your school has been closed for the remainder of the term. You can't hold any in-person classes or club activities. The spring musical is canceled, and sports are delayed indefinitely.

Your students don't return from their break, and you're left with no option but to teach online with only a few days' notice.

This scenario sounds like a horror film written by a high school science teacher, but it's the reality that many K-12 educators experienced in the spring of 2020. When the COVID-19 pandemic hit, instructors from preschool to postgraduate were forced to pivot to online instruction. Even in a time when flipped, hybrid, and entirely online classes aren't novel, teaching with technology presented new challenges—and opportunities—for educators across the globe.

Objectives

By the end of this chapter, learners will be able to:

- Understand how technology can be used for content delivery, classroom management, and student assessment.
- Evaluate platforms and their suitability for different instructional situations.
- Design learning opportunities that utilize technology for content delivery, classroom management, and assessment.

Introduction

In this chapter, we'll discuss some of the basic tenets of content delivery and incorporating technology in your classroom. According to the Council for the Accreditation of Educator Preparation (CAEP), preservice teachers should be able to use technology in content delivery, classroom management, and student assessment. We'll explore how to apply those three goals to a variety of technologies and platforms.

Overview

Using technology for the sake of using technology is no longer an option. Educators must routinely evaluate how technology can assist them in delivering content, engaging diverse learners, supporting and assessing student learning, and empowering students to connect and collaborate in a hyperconnected world. While many frameworks for technology integration exist, TPACK (see Koehler & Mishra, 2009) is a robust framework that focuses on the types of knowledge a teacher must possess to integrate technology efficiently and effectively.

Technological Pedagogical Content Knowledge

Technological Pedagogical Content Knowledge (TPACK) is a conceptual model that seeks to identify the complexities of knowledge required by teachers integrating technology. The three primary forms of knowledge include Content Knowledge (CK), Pedagogical Knowledge (PK), and Technological Knowledge (TK). The types of knowledge where each of those forms intersect include Pedagogical Content Knowledge (PCK), Technological Content Knowledge (TCK), Technological Pedagogical Knowledge (TPK), and Technological Pedagogical Content Knowledge (TPACK) (see figure 8.1). These relationships are complex, dynamic, and transactional. "Individual teachers, grade-level, school-specific factors, demographics, culture, and other factors ensure that every situation is unique, and no single combination of content, technology, and pedagogy will apply for every teacher, every course, or every view of teaching" (Koehler, 2012, para. 3).

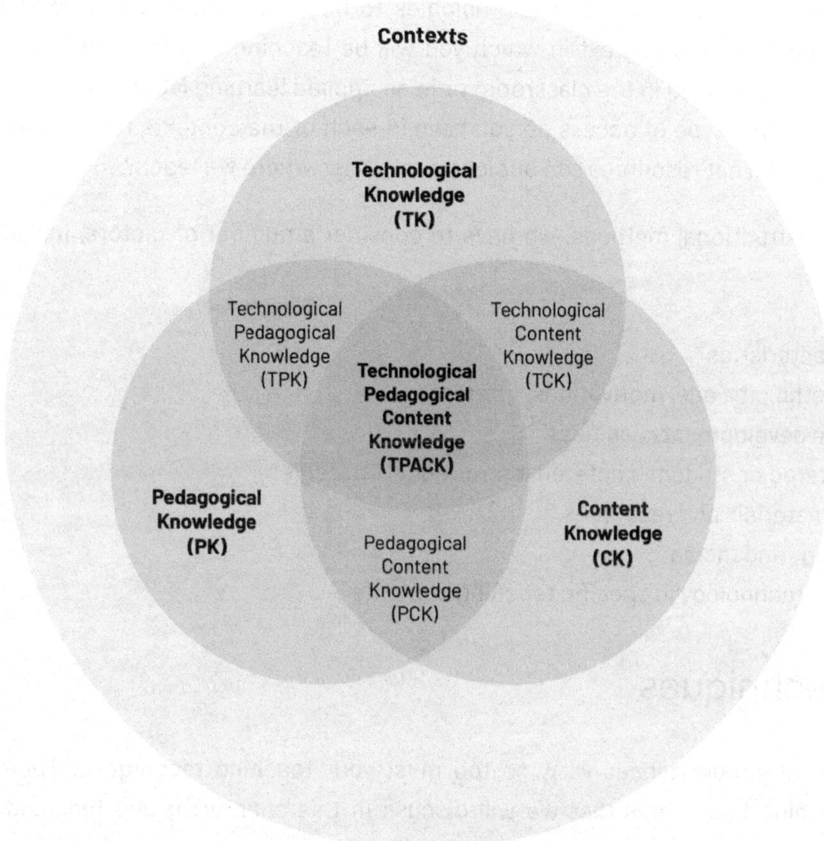

Figure 8.1: Technological pedagogical content knowledge (TPACK).

Content Delivery

Teaching Techniques/Instructional Methods

After establishing learning objectives, educators must determine the teaching technique(s)—essentially, how we choose to deliver the planned content (as discussed in the previous chapter)—that will be most appropriate for ensuring students achieve the objectives.

There are many factors to consider when selecting the teaching techniques, materials, and resources that you will need to teach effectively and efficiently. This chapter focuses on specific teaching techniques and provides an overview of the types of technology that you should be familiar with to plan and deliver content. We also recommend reviewing the chapters in this textbook on diversity, equity, and inclusion and inclusive teaching before you continue!

When selecting teaching techniques and technologies to be integrated into your learning environment, it is important to consider the context in which you will be teaching as well as student preferences and capacities. Will you be teaching in the classroom or in an applied learning laboratory (i.e., greenhouse, land lab, ag mech shop)? What type of access do you have in each of the contexts that you may teach in as an agriculture instructor? What resources can our learners access where we teach?

When we choose instructional methods, we have to consider a number of factors, including our learners' characteristics.

- Learner characteristics
    - Gender, ethnicity, age, motivations, interests
    - Cognitive development readiness
- Teacher-centered or student-centered instructional strategies
- Selection of materials and resources
    - Technology and media
    - Access to technology in specific teaching contexts

Teaching Techniques

As your instructional circumstances vary, so too must your teaching techniques. There are two broad categories of teaching techniques that we will discuss in this chapter: group teaching techniques and individual teaching techniques. As an educator, it will be important for you to choose the appropriate technique(s) for a given lesson. You must consider the subject matter to be taught, the objectives to be achieved, learner preferences, teacher preferences, and resources available. As you ponder each of these, there will likely be obvious teaching techniques that best fit your situation. We will explore each of the categories and define specific techniques within each in more detail below.

Group Teaching Techniques

Group teaching techniques are useful when providing instruction to learners in the same setting. If your learning environment is conducive for group teaching techniques, you can maximize your efforts in instructing students that have common needs. Group teaching techniques are appropriate when considering the limited time educators have with learners and when considering that often in educational settings, learners should all have some common factual information for progressing through a lesson. Group teaching techniques are also useful in managing the potential lack of resources available. For example, if a learning environment is short on student reference materials or electronic devices, group teaching may be the most efficient method for educators to teach learners the information needed for a given lesson.

Basic Group Teaching Techniques

Seven commonly used group teaching techniques are:

1. Lecture
2. Discussion
3. Demonstrations
4. Field trips
5. Role-play
6. Resource people
7. Cooperative learning

Each technique will be explained in further detail. Table 8.1 provides a brief description of each technique and how an educator may incorporate technology.

Lecture

The lecture is the most used group teaching technique and is useful for sharing factual information. It may often have negative connotations as teachers often fail to plan for lectures that adequately engage learners. Organized lectures with appropriate summaries and clear conclusions are vital for student learning. Visuals are also a useful tool to help learners develop mental images that help provide important frames of reference. Lack of enthusiasm is a major hurdle that educators must overcome when using lectures. How can you effectively use lectures with your learners? Planning is crucial. You should develop a thorough outline of the content to be taught and important concepts to highlight. The organization should be clear and concise and make use of appropriate visuals to help students create mental images. You may choose to organize your lecture outline in any of several orders, such as

- Chronological
- Spatial
- Cause to effect or effect to cause
- Problems and solutions (problems, possible solutions to each, conclusions for each)
- Topical

As you present your lecture, you should be enthusiastic and avoid distracting mannerisms. Leverage visuals to aid in illustrating the major points and provide summaries throughout the lecture and again at the end. The end of the lecture should often present very clear and specific conclusions.

Discussion

Discussions are often used when the educator wishes to serve as a facilitator while learners develop critical thinking and argumentation skills. Discussions require careful planning as leading questions are often the driver of this teaching technique. The important categories of information to include in a discussion plan are:

1. Sequence
2. Important subject matter and key points to highlight
3. Leading questions
4. Teacher directions

As you prepare your discussion plan, the sequence of events is critical to ensure it has a logical flow, thereby making it easier for students to follow. Failure to include the important subject matter or key points to highlight will likely result in the omission of that information. Having the important points documented in your plan serves as a reminder to emphasize the information during the discussion. For new and beginning educators, questioning can be a difficult teaching skill. You will want to spend time developing questions to help your learners think critically about the subject matter being learned. Teacher directions serve as a guide for specific examples that you want to share with your learners during the discussion. For example, you may include a note to share a specific example to help clarify a point during the discussion. Altogether, these four categories of information can help ensure you facilitate an effective and engaging discussion.

Another important aspect to consider when planning discussions is the specific format you will use. Common formats include whole class, brainstorming, buzz groups, and pair-share.

Whole Class

Leading questions are posed, and learners offer answers. This process continues until the major points to be covered are clearly explained. The educator serves as the facilitator of the discussion to ensure it stays on track and progresses appropriately.

Brainstorming

You can also provide the whole class or small groups of learners with a specific problem or topic and ask them to develop as many ideas, answers, or solutions as they are able to. Quantity versus quality is usually desired when utilizing the brainstorming discussion method. As such, it is important to not hamper creativity or judge student responses. Once the list of responses has been gathered, they can be evaluated and pared down as needed to meet the objectives of the lesson.

Buzz Groups

Buzz groups are small groups of three to seven learners that discuss a specific problem or topic and develop the best possible solution or answers. In contrast to brainstorming, which is concerned with quantity, buzz groups are more focused and purposeful. After the individual groups have finished the discussion, the educator facilitates reporting out to the whole class.

Pair-Share

Pair-Share is a quick and effective discussion method that tasks learners with developing their own idea or opinion to a question or problem before pairing with another learning and sharing. The two learners discuss and try to formulate the best solution collaboratively.

Discussion can be intimidating to learners for myriad reasons. Topic selection, fear of ridicule, or ambiguous questions are just a few examples. A skillful educator will plan discussions in such a way that learners feel comfortable engaging with their peers.

Demonstrations

The psychomotor domain is developed through hands-on application. Agriculture, with its many psychomotor skills that need to be practiced, is bursting with opportunities for demonstrations. Simply talking about a skill will not translate to the learner acquiring the ability. Learners need to see and practice to gain specific psychomotor skills. As the teacher, it is your responsibility to produce a step-by-step procedure that learners can follow. You must be prepared with explanations, illustrations, and provide the learners with an opportunity to practice. When planning to use demonstration, you must determine what you wish to accomplish, which must align with what is possible. This means the teaching objective must be able to be demonstrated, have an appropriate scope, and be appropriately challenging.

While spontaneous demonstrations can arise in a learning situation, most should be planned. Once you have selected a skill that would be best taught via demonstration, develop the step-by-step instructions to follow. Be sure to indicate the key points to share for each step. Then you should create a list of the necessary materials for each step in the demonstration. This is crucial and helps avoid oversight during the initial and succeeding demonstrations. As you prepare to teach the selected skill, following these simple steps will ensure a successful demonstration:

1. Assemble materials
2. Establish the "What's in it for me?" on the part of the learner
3. Provide an overview of the process you will be demonstrating
4. Present each step
5. Repeat challenging steps as necessary
6. Repeat the entire process
7. Allow learners to practice

The range of skills that an agricultural educator may want to demonstrate is vast. To ensure a successful demonstration, you must possess a certain level of mastery for the skill being demonstrated. This may require practice on your part or the selection of a different teaching technique. Total mastery on your part may cause you to omit certain steps that a novice may need to be successful. You must think about where your learners are at and plan accordingly. Another concern is class size. Large class sizes can be a hindering factor in demonstrations as some learners may not be able to see or hear you walking them through the steps. Supervision of learners practicing is also a factor you need to plan for. You may need to develop a rotation procedure to allow learners an opportunity to practice the skill. Safety should always be considered.

Field Trips

A field trip is a great way to provide students with opportunities to observe and participate in real-world situations. While your program may not have the facilities or resources to accompany instruction in agriculture subject matter, you can make use of field trips to provide concrete frames of reference for students. Field trips are a great way to vary your instruction and provide an enriching learning experience for students. Field trips can be used to create interest at the beginning of a unit, to provide closure and reinforce the relevance of the content at the end of a unit, and for deeper exploration of what is being learned anytime in between.

Effective field trips require careful planning. Failure to plan will certainly lead to a less than desirable experience for all involved. Follow these steps to ensure your field trip is successful.

1. Establish the goals for the field trip.
    a. Determine the specific learning objectives that the trip will address.
2. Select an appropriate site for the field trip.
    a. Consider the quality of the experience to be seen, a site's willingness to host learners, the distance to the site, and the cost.
3. Secure approval from your administrator
    a. Follow your local policies and procedures for requesting field trips. Some school districts may require advance notice of a month or more.
4. Make specific arrangements for the field trip.
    a. Select a date and time, create an outline of what the learners need to see and do to align with the goals and objectives of the trip, and always check on the parking situation (especially if you are traveling by bus).
5. Prepare students for the field trip.
    a. Work with your learners to develop a set of questions to be answered by the field trip. Communicate the objectives of the experience to the learners so they know what to expect when they return. Discuss proper conduct, expectations for attire, eating arrangements, and any money needed for the trip.
6. Conduct the field trip.
    a. Ensure you have all the necessary documentation based on local policy and procedures. Provide the host site with a courtesy reminder about your arrival time. Manage student behavior.
7. Summarize and reflect on the field trip experience.
    a. The field trip should not end upon departure from the field trip site. You should facilitate a summary and reflection exercise with the learners to develop conclusions to be drawn from the experience. Challenge your learners to develop plans for applying what they have learned.

Role-play

Providing learners an opportunity to practice specific skills or abilities in an approximated real-world scenario can be a powerful experience. As you consider the skills that learners need for competing in society, using the role-playing teaching technique may provide them with an edge. Role-playing involves having those in the learning environment play or portray a given role. The experience of playing the role and the analysis of the situation help learners develop the skills or information needed for the subject being learned. Do you want your learners to master basic agricultural sales skills? Should they know how to properly introduce strangers or conduct a phone conversation with important stakeholders? This is where role-playing shines. It is most effective for topics relating to human relations, leadership, and sales skills.

When deciding to use role-playing, ensure the situation and assessment are aligned to the objectives. Meaningful closure should be provided to capture the key learnings to be gleaned from the role-play. Since the role-play experience will involve lots of student-student interaction, you should be explicit in laying out your expectations. Communicate the purpose of the role-play with your learners and the importance of supporting each other. Being teased or ridiculed often becomes a barrier to learning.

Resource People

Resource people are incredibly useful in teaching agriculture. You are likely not an expert in all subjects you may teach, but as an effective educator, you know how to ensure your learners are getting the knowledge they need to be successful. Resource people are often experts and can help deliver needed information or provide a different perspective on a topic. Using resource people helps build connections between your learners and key individuals in the community—those that may help your learners make decisions in the future. To be most effective, resource people should be more than guest speakers. Resource people may supplement what is being learned, help add relevance to the topic, or endorse a specific process or practice.

Resource people are akin to consultants to the agricultural educator. As such, you should ensure that you provide the individual with information about your learners and the specific objectives for what is being taught. Work with your learners prior to introducing the resource person to develop questions to maximize the effectiveness of this teaching technique. You should be actively involved in any interaction between the resource people and your learners. For example, you may need to promote the raising of questions from learners or provide clarity on the information presented by the resource person. You want resource people to have a positive experience with your learners, so you need to communicate expectations and encourage appropriate behavior for interaction.

Cooperative Learning

Cooperative learning is a teaching technique that empowers students to assume more responsibility for their own learning. Cooperative learning promotes the development of group processing and social skills, which are important college and/or career readiness skills. Learners, in groups of three to five, work together toward a common learning goal. The groups are often intentionally selected, and the learners assign roles once they begin working together. Depending on the situation, these cooperative groups may work together for one lesson or several weeks. Learners help hold each other accountable and provide support as they work toward the learning goal.

Technique	Description	Incorporating Technology
Lecture	Transmit factual information, provide explanations, supplement or enhance readings	Record short lecture videos or podcasts
Discussion • Whole class • Brainstorm • Buzz group • Pair-share	Students respond to a well thought out, leading question until the major points have been developed and explained	Create an online discussion board with the class or others from around the world
Demonstrations	Show students how to perform a process or task	Record video demonstrations, VR demonstrations
Field Trips	Showcase a real-world setting related to course content	Conduct virtual field trips, VR tours
Role-Play	Portray a given role and analyze it to learn about a topic or concepts	Act out a sales pitch to a potential client via videoconferencing platforms
Resource People	Work with a "consultant" who aids in teaching or reinforcing content	Connect with experts from around the world via videoconferencing platforms
Cooperative Learning	Intentionally select groups of three to five learners to work together on a well-defined learning task	Utilize a digital collaboration tool to facilitate group work

Table 8.1: Group teaching techniques.

Individualized Teaching Techniques

Every learner has their own unique needs. In an ideal situation, the teacher would establish specific goals for each course and every learner could work to achieve those goals consistent with their own unique needs and abilities. Each learner would have the needed materials, approaches, and support to assist them on their learning journey. In formal settings, this ideal situation can be difficult. However, a quality educator remains committed to supporting every learner and can do so by leveraging individualized teaching techniques. Great educators will use a variety of group teaching techniques and individualized teaching techniques. This provides variety to the learning activities. As you consider individualized teaching techniques, you must be aware of the needs of your learners. These techniques not only allow you to individualize the learning, but they can also help learners develop the ability to inquire into subject matter. Learners no longer need to wait for someone to tell them what they need to know, as they are actively seeking information. This helps them become independent problem solvers. Through the selection of appropriate individualized teaching techniques, learners begin to critically evaluate and apply information. Leveraging individualized teaching techniques can empower learners to take ownership of their learning journey.

Basic Individualized Teaching Techniques

Five commonly used individualized teaching techniques are:

1. Supervised study
2. Experiments
3. Independent study
4. Student notebooks
5. Information sheets, assignment sheets, skill sheets

Each technique will be explained in further detail. Table 8.3 provides a brief description of each technique and how an educator may incorporate technology.

Supervised Study

Supervised study is a technique that allows learners to use basic reference materials to find answers for themselves. The learners do not have to depend on the teacher to provide the information and can obtain specific information that they need independently from what other learners may need. Supervised study also allows the teacher to work with individuals that may need additional support. As with any teaching technique, planning is crucial. You should determine the desired outcome of the supervised study, decide on the resources to be used, and plan how the learners will conduct the supervised study. There are six basic types of supervised study which include:

1. All learners study the same problem using the same reference materials.
2. All learners study the same problem using different reference materials.
3. Small groups study the same problem using the same reference materials.
4. Small groups study the same problem using different reference materials.
5. Small groups study different problems using the same reference materials.
6. Small groups study different problems using different reference materials.

The decision on which type of supervised study is based on the reference materials available and the desired outcomes of the technique. It is important that prior to implementing supervised study, you clearly define the problem, communicate what information the learners need to discover, and provide direction to meet the desired outcomes.

Experiments

Experiments create a high level of learner involvement, which often translates to increased interest. Experiments challenge students to use their minds and to be physically involved in setting up the experiment, making observations, collecting data, and developing conclusions. Experiments allow for concepts, theories, or approved practices to be graphically illustrated. This teaching technique helps promote systematic thinking. You must emphasize the importance of being systematic, clear, and thorough throughout the experiment. In drawing conclusions, learners learn that all the facts must be considered.

As you consider using this technique, you can identify an appropriate experiment or work with your learners to design an experiment to solve a problem. Once the experiment is identified, outline the procedures to follow, how to collect and record data, and a list of supplies and materials needed. If you design an experiment for learners to conduct to solve a problem, it is recommended that you also have an idea of the basic findings and conclusions the experiment is intended to reveal. At the conclusion of the experiment, the whole group should benefit from the knowledge gained, which is generally achieved through a report. You should work with your learners to standardize reporting procedures. Table 8.2 offers a list of ideas for experiments. Though not exhaustive, it is intended to help you begin thinking of other ideas for experiments.

Subject Area	Ideas for Experiments
Production Agriculture	• What herbicide controls weeds best? • Which ration is best for a given species of livestock? • How does planting date affect yield?
Horticulture	• What is the most effective way to propagate a given plant? • What effect does soil sterilization have on plant growth? • What is the best method to control a given insect or disease?
Agricultural Mechanics	• How does temperature affect oil viscosity? • How does the spark plug gap size affect engine starting performance? • How does amperage affect weld strength?
At-Home Experiments	• How does technology use before bed impact sleep quality? • How does the consumption of different types of food at breakfast affect energy levels throughout the morning? • How does exposure to natural light versus artificial light affect overall mood?

Table 8.2: Ideas for experiments by subject area.

Independent Study

Independent study is any form of study that is conducted by an individual learner. Supervised study and experiments, if completed by individual learners, are considered independent studies. However, the independent study can be leveraged beyond those examples. The main goal of independent study is to meet the needs on individual learners. You may encounter a situation where a learner can't physically complete a planned learning activity. Therefore, independent study can provide a meaningful alternative. Independent study can be reading, self-paced instructional units, computer-assisted instruction, or many other formats.

During independent study, you take on a facilitator role. The learner assumes responsibility for their own learning. This technique can add variety to your instruction, but it also promotes learner independence. You may need to teach learners how to study independently for this technique to be successful. Additionally, you should work with your learners to apply what is learned from the independent study so that it is not a meaningless exercise. Independent study can occur during class or lab or outside of traditional instructional time. While the format for each learner's independent study will vary, some structure is still needed to help them progress. A sample structure is below.

Sample Independent Study Structure

- What is the problem to be solved?
- What are the questions that must be answered?
- What references will I use?
- What steps will I follow in conducting my independent study?
- What will I learn from my study (a summary)?

Notebooks

Notebooks are a great way for learners to organize and accumulate their knowledge and can serve as a constant reminder of what was learned and how all the pieces fit together within a unit or course. Notebooks can also be a great reference for future problems or studying for quizzes or tests. Good note keeping promotes learning. High-quality notebooks require a commitment from the educator. You have to teach your learners how to set up their notebooks and the expectations for adding to the notebook as learning progresses. You can provide learners with a recommended note-taking format to keep everything uniform. You should continually remind learners of the importance of a high-quality notebook and evaluate the notebooks periodically. The format, electronic or hard copy, can be based on your preference or the needs of your learners.

Study Sheets

Another way to individualize learning is to use specific study sheets that are designed to guide a learning experience. Information sheets, assignment sheets, and skill sheets are the common study sheets used and are most often associated with laboratory learning. These study sheets allow learners to learn at their own pace and to progress when they are ready to do so. The study sheets are narrow in scope and should be developed to be suitable for the majority of the learners.

Information Sheets

Information sheets are short handouts that include basic information (the what, why, and how) a learner needs in order to perform a specific skill or job. While they can be used to introduce learners to information for the first time, they are also useful in providing a synopsis or summary of previously covered material.

Assignment Sheets

An assignment sheet details an assignment that learners should complete, how to complete it, and how to check the assignment for completeness.

Skill Sheets

Skill sheets are similar to assignment sheets but are more limited in scope. Skill sheets focus on a specific skill and provide guidance to the learner in developing and mastering the skill. Step-by-step instructions with illustrations help learners visualize the specific skill.

Many factors go into considering the teaching techniques to be used. For individualized teaching techniques, it is imperative that you consider the readiness of the learner. In order for a learner to pursue independent study, they must be able to organize thoughts, read, manage personal behavior, and synthesize information. You may have some learners that will never fully meet these conditions while others may thrive. This highlights the importance of knowing the needs, skills, abilities, and interests of your learners. Variety is important and with the range of group and individual teaching techniques discussed, you can maintain student interest while working toward the desired outcomes of your program.

Technique	Description	Incorporating Technology
Supervised Study	Seek answers to clearly defined problems with guidance from the teachers (i.e., providing specific resources). Work with the students to arrive at a final conclusion with this method.	Use the internet to search for the information being sought; Use online user manuals to research information based on questions developed by the teacher
Experiments	Explore a phenomenon (most often via the scientific method). Gather facts and draw conclusions and graphically illustrate important concepts, theories, or practices.	Electronically log of all the steps in an experiment; Digitally report the results of an experiment; Use electronic sensors to help conduct experiments
Independent Study	Learners individually conduct a study. Use this opportunity to meet their individual needs.	Use an electronic form to gather information from learners about their interests and abilities related to the course to assist in designing independent study tasks; Allow learners to use the internet or other digital resources for independent study
Student Notebooks	Organize and document learning throughout a course.	Allow learners to create a digital notebook to document their learning; Use digital interactive notebooks
Information Sheets, Assignment Sheets, Skill Sheets	Use specific study sheets that are designed to aid in guiding learning experiences.	Use digital study sheets; Use video information or assignment sheets

Table 8.3: Individual teaching techniques.

Incorporating Technology in Instruction

Technology has transformed many aspects of our lives. It has the potential to do the same for teaching and learning. As educators, we should embrace and leverage the transformative power of technology to create exciting and engaging learning opportunities, or risk leaving our students at a competitive disadvantage in the technologically advanced workforce. Technology integration requires careful planning and appropriate pedagogical practices to be successful.

Common forms of technology we incorporate into our learning environments include productivity software, learning management systems, content delivery platforms, and assessment tools. Generally speaking, we will likely use technology to plan, create, and deliver content; assess learning; and empower students as creators.

Examples of technology used to plan, create, and deliver content and to assess learning may include:

- Productivity software (e.g., Microsoft Word, Google Docs), computer, projector, smartboard
- Presentation software (e.g., Microsoft PowerPoint, Google Slides, Prezi)
- Graphics/Videos
- Learning management systems (e.g., Canvas, Google Classroom)

Examples of technology we may use to empower students to create high-quality products or representations of their learning may include:

- Web 2.0 tools (e.g., blogs, social media)
- Presentations, graphics, videos

Table 8.4, although not exhaustive, provides a couple of easy-to-implement examples of how to leverage technology from a teacher and learner perspective.

Technology Example	Teacher Use	Learner Use
Productivity Software	Use a word processor to develop unit and lesson plans.	Use presentation software to create a project report.
Presentation Software	Use a video recording platform to record short lectures or instructions for learning centers.	Have students develop video presentations to represent their learning.

Table 8.4: How to leverage technology.

Digital Divide, Access, and Accessibility

Although technology has impacted many aspects of our daily lives, its full potential remains out of reach to millions of people. The term "digital divide" refers to the gap between individuals who have access to computers and the internet and those who don't.

It is important to analyze our specific situations to know what is and isn't appropriate in our local communities. Access to broadband internet remains a privilege with very limited access to individuals in rural areas. According to the Federal Communications Commission, a quarter of rural Americans and Americans who live in Tribal lands lack access to reliable and affordable high-speed internet. Lower-income families are less likely to own personal computers or tablets. Many learners rely on school programs to provide devices for educational use.

As we prepare for instruction, we should consider what we already have. What technology resources are available in our learning environments, our schools, our communities? What access do our learners have and what access do they need? Develop a short survey asking about learners' technology access and use. Develop opportunities for learners of all access levels. For example, post instructional materials online, but also have physical copies for learners without access.

There are also many governmental and nongovernmental programs to help fund technology-assisted learning. The United States Office of Educational Technology, part of the US Department of Education, offers grants and funds to support remote learning and connectivity (for more information, visit https://tech.ed.gov/funding/). We can also leverage other creative solutions, such as crowdsourcing and fundraising, to support technology in our learning environments.

Accessibility and Assistive Technology

Accessibility is a broad concept that applies to content, devices, and digital tools. Accessible technology removes barriers for students with varying cognitive and physical abilities. As educators, we have a moral and legal obligation to ensure that we provide our learners with accessible materials, including content and technological tools.

Some forms of technology have been specifically created to help learners fully integrate into a learning environment. Assistive technology is equipment or software that helps learners navigate their environments while managing challenges with learning, communication, or mobility. Assistive technology includes high-tech tools like augmentative communication devices or accessibility features on devices such as text to speech, word prediction, and optical character recognition. The term can also apply to physical parts of the learning environment, including furniture and peripheral tools like keyboards.

Examples of accessible content may include closed captions for videos, alt-text that describes an image, text-to-speech functionality, or screen magnifiers. Most technological tools have built-in accessibility checkers and can provide recommendations for changes that can make content more accessible. Additionally, we can work with resource personnel to ensure the technology we use or the content we create is accessible to all learners, regardless of abilities. For more information, visit https://www.w3.org/TR/WCAG20/.

Specific Examples of Technology Integration

Learning Management Systems

A learning management system (LMS) is a set of software that centralizes the administration, tracking, and reporting of educational activities. An LMS allows instructors to develop a course schedule, register students, and assess their learning outcomes in one platform (Ninoriya et al., 2011). LMSs are used for a variety of administrative and teaching tasks, including tracking student attendance, delivering content, and recording assessments.

The list of enterprise (paid) and free LMS options are always evolving. At the time of this printing, the top three enterprise LMSs used in the United States are Blackboard, Canvas, and Moodle (Edutechnica, 2020). Google Classroom is a free alternative that integrates email, calendars, and document sharing within the Google app suite; by 2017, it was used by half of K-12 students in the United States (Singer, 2017). Schoology incorporates elements of social media and gamification—applying elements of game playing, like competition and scoring points—to educational activities in its interface, making it well suited for all learners (Fenton, 2017).

Learning management systems have many benefits. They are portable, often accessible from a host of mobile devices, and scalable, meaning they can be implemented in both small and large groups of learners. They can be integrated into hybrid instruction, or teaching that includes elements of both in-person and distance learning (Whiteside, Dikkers, & Lewis, 2017), allowing instructors to move from a physical learning environment into the digital space.

At the same time, it's important to recognize LMS platforms' drawbacks: They can be expensive to license, especially if they're not supported by a school system, though free options exist. They require technology inputs, including stable internet connections and hardware like laptops or tablets. There is also something of a training learning curve for instructors, especially as technologies and platforms are constantly changing. Care must also be taken to ensure that the LMS and its tools and operations are accessible to users with physical or cognitive challenges.

Strategies and Best Practices for Teaching, Management, and Assessment with an LMS

When teaching with an LMS, you must remember that it is not itself a teaching tool; rather, it's a means of organizing and sharing information and activities with learners. There are a number of web tools available that can be incorporated into and deployed by an LMS, but we shouldn't rely on the LMS alone to deliver content.

When developing content to share via LMS, remember to take advantage of multimedia features. For example, a lecture captured on video may be more engaging than a list of slides for students to review. Many LMS platforms allow learners to upload responses in audiovisual format, turning them from multimedia consumers into content creators.

From a management standpoint, it's important to set expectations for learner and instructor activities in the LMS. Before we begin a session, we should explain how our LMS works, and how we will use it in our learning environments. Make sure learners are comfortable with the format and functions before turning them loose.

LMS platforms allow us to assess learning and gather feedback through quizzes, surveys, and polls. These assessments let instructors choose from a variety of question types, set time limits and deadlines, and ease their workload through automatic grading of multiple-choice, matching, or true/false items where applicable. Learners can also be prompted to upload portions of an assessment—for example, a schematic they've drawn or calculations they've attempted.

Teaching Techniques Reimagined

Virtual Field Trips

Firsthand experiences like field trips are important parts of an educational experience. However, visiting locations such as farms, production facilities, and factories is becoming increasingly difficult due to tightening biosecurity, health, and safety restrictions. In addition, in-person visits require funding and instructor time to coordinate schedules, acquire approvals, and arrange transportation. Virtual field trips (VFTs) offer an alternative for getting learners into off-limits or geographically distant spaces.

Like in-person field trips, VFTs help learners understand and connect with a profession by placing them in the field with experts—the key difference being that VFTs rely on technology, like prerecorded videos, 360-degree viewers, and teleconferencing software, to make the "visit" happen. VFTs may be synchronous, meaning that learners and hosts interact with each other live via web cameras and microphones, or asynchronous, meaning that the virtual tour or visit is recorded in advance and shared in an online platform like YouTube or an organization website.

VFT Best Practices

When we're considering participating in or creating a VFT, there are several elements to consider. It's important to introduce our students to the subject of the experience before engaging in the virtual trip. These pre-VFT materials could include readings, pretests, and in-class activities. As with any learning endeavor, we should also be prepared to assess student engagement and learning after the experience. Posttests, surveys, or debrief discussions can help reinforce key takeaways. We can incorporate a VFT into a project that uses information gained in the field trip, or even assign students to plan and host their own virtual trips with instructor support.

Preparation is key to a successful virtual field trip. Hosts in the field and instructors in the class should all test their equipment, including cameras, microphones, and internet connections, in advance of the experience. If the VFT is synchronous, ask learners to prepare questions for the speaker or host in advance. Though new questions will almost certainly arise as the VFT ensues, this will prevent any pauses or disruptions to the flow of conversation. If we're planning or organizing a VFT, recording and archiving these experiences, so that learners can refer back to them, is also helpful.

Immersive Technology

Immersive technology is a term used to describe technology that replaces or extends the physical environment. Examples include augmented and virtual reality. Augmented reality (AR) is when digital artifacts are overlaid into a real environment on a screen (e.g., smartphone). Virtual reality (VR) is typically defined as a fully immersive experience where the environment is completely replaced with a digital environment, usually via a headset (see figure 8.2).

Figure 8.2: Tethered, stand-alone, and mobile VR headsets.

Virtual Reality

Originally, fully immersive VR experiences required a powerful VR-ready computer and a tethered VR headset. Now, stand-alone VR headsets that don't require a costly computer are readily available. Costly computer equipment can be a barrier for many schools. The power of VR is still within reach through mobile VR experiences. Mobile VR utilizes smartphones to deliver limited-capability VR experiences. These experiences are most often in the form of 360-degree videos that can be used with or without a mobile VR viewer.

Augmented Reality

Augmented reality brings abstract concepts to life via a smartphone or tablet. Students can now hold a strand of DNA or rapidly prototype and see their creations as a digital artifact before production. AR can also provide information on objects that can enhance and extend learning.

Immersive experiences are well supported with several pedagogical theories. Such theories that support immersive tech and VR use include situated learning, experiential learning, embodied interaction. (For more information about pedagogical theories see chapter 3, "Principles of Teaching and Learning.")

Much like VFTs, immersive technology applications can mitigate safety and security hazards that are associated with agricultural occupations. You can use immersive experiences to train students, to gamify content, or to connect students to experts in other parts of the world. Immersive experiences can eliminate time, place, space, and financial constraints.

We have used immersive experiences to help train students on shop safety, candling eggs, proper cattle handling, veterinary skills training, and animal waste management.

Virtual Simulations

Simulations have long been a staple in educational settings. Simulations put students into situations that mimic the real world in order to create a real-feel experience. Many professions rely on simulations as they offer lifelike experiences for beginners and experienced individuals alike. Depending on the complexity and the technology readily available in your classroom or school, simulations can be used for an entire class, groups of students, or as an individual activity.

Consuming versus Creating

Many immersive applications exist that can be used in your program, but to take learning to the next level, you can shift students from consumers of those experiences to creators of those experiences. For instance, you can have students create their own virtual tours or immersive training content with 360-degree cameras (or a smartphone application) and readily available software.

Example Checklist for Creating Virtual Tours

- Identify the purpose of the virtual tour.
  - How does the virtual tour help students achieve the objectives?
  - What do you want the students to gain from the virtual tour?
- Create a storyboard.
  - Sketch out the scenes or develop a list of scenes that contain what you want students to experience.
  - Include points of interest that you want students to notice in each scene.
  - Write out instructions for each scene.
- Visit the site to take 360 photos (or use existing 360 photos).
  - Using a 360 camera, or existing 360 photos, collect all the needed assets.
- Create the virtual tour.
  - Using your preferred virtual tour platform, construct the virtual tour based on your storyboard.

Many agricultural communications programs create publications, public relations campaigns, and other communications artifacts as projects in advanced courses that can be shared with a wide audience. They assume roles (remember role-playing?) as staff members for a magazine or communications firm, utilize their skills, and produce materials that they can distribute to clients, readers, or other stakeholders. Publications like newsletters and magazines can be easily shared with inexpensive tools such as Issuu, an online publishing platform. Educators can get free licenses for tools like Adobe Express, which allows students to create social media posts, interactive web pages, and videos using tablets or smartphones. Students at Ohio State University have created many great examples of this type of project. To see their work, visit https://u.osu.edu/agrinaturalist/.

Generative Artificial Intelligence

Generative Artificial Intelligence (AI) is a subset of artificial intelligence that leverages machine learning models to generate content. These models are trained on large language data sets, learning to create similar outputs based on the patterns and structures they've learned. This includes creating anything from articles and reports to poems, music, and even realistic images. Generative AI has grown in popularity and is becoming ubiquitous across many industries, including education.

Pros of Generative AI in Education

One of the most powerful uses of generative AI in education is the potential for personalized learning. By analyzing a student's learning patterns, comprehension level, and areas of strength and weaknesses, generative AI can create customized educational content tailored specifically to that student's needs. This enhances the learning experience and can lead to improved academic performance.

Generative AI also has immense potential in making education more accessible. For instance, it can be used to generate learning materials in multiple languages or create synthesized speech for visually impaired students.

Furthermore, generative AI can help educators by automating some of their workload. It can generate quizzes, assignments, and tests based on a given syllabus or learning material, freeing up time for educators to focus on instruction and student engagement.

Finally, generative AI can be used as a creative tool in the classroom. Students could use it to generate ideas for essays, stories, or art projects. It's a great way to introduce students to AI technology and inspire creativity.

Cons of Generative AI in Education

While generative AI holds great potential, it's essential to be aware of its limitations and challenges. One of the biggest concerns is the quality and accuracy of generated content. Without careful oversight, AI can produce content that is misleading or inaccurate.

Generative AI also presents ethical considerations. These systems could potentially be used to create plagiarized work or disinformation, so educators must instill ethical usage principles in their students.

Furthermore, there's the risk of over-reliance on AI in the learning process. Generative AI should not replace the role of educators but rather serve as a tool to supplement and enrich the educational experience.

Examples of Generative AI in the Classroom

Chatbots powered by generative AI can simulate conversations with students, answering queries and providing explanations about a particular topic. Generative AI can be used in storytelling or creative writing classes, where a model like GPT-4 can generate story prompts or continue a narrative based on student input. This can be an exciting tool to stimulate creative thinking and improve writing skills.

As you promote inquiry among your learners, generative AI can be leveraged to help brainstorm and refine research questions for projects or experiments. It can even be used to provide suggestions for methodologies or sources of information.

While exploring history, generative AI can create simulations or scenarios based on historical events, providing students with a more immersive and engaging way to learn about the past.

While generative AI offers numerous exciting opportunities for teaching and learning, it's important to understand its limitations and employ it responsibly. The efficacy of using generative AI in education is when it's used as a tool to supplement, rather than replace, traditional teaching methods. The future of education is likely to be a blend of human and AI-driven teaching methods, offering a more personalized, engaging, and effective learning experience.

Sample Prompts for Generative AI

- "Can you describe what a career in environmental science might look like?"
- "Develop a fifty-minute lesson plan for ninth graders on the basics of floral design. Include an interest approach, two different teaching techniques, and a check for understanding."

Other Topics to Consider

- **Copyright.** Copyright is a form of intellectual property protection for authors of published and unpublished works. As an educator, you have a legal and ethical obligation to respect copyrighted material. You serve as a model for your students in doing so.
- **Fair Use.** Fair use is a legal doctrine and framework for making a determination on whether or not using certain material is a copyright infringement. Fair use is commonly cited for nonprofit educational purposes.
- **Acceptable Use.** Set of rules established by the owner of a computer or network that defines the guidelines as to how it should be used.

Learning Confirmation

1. Define the following key terms.
    a. TPACK
    b. Digital divide
    c. Accessibility
    d. Learning management system
    e. Hybrid instruction
    f. Synchronous instruction
    g. Asynchronous instruction
2. Develop a technology matrix with ten different tools/platforms. In your matrix, provide a brief description of the tool and identify how you would use it for content delivery, classroom management, or student assessment.
3. Your travel budget has been limited for the second year in a row. How could you offer your learners an opportunity to visit a desired destination to extend the learning for a unit/lesson in your curriculum?

Applying the Content

You are teaching in a secondary school-based agricultural education (SBAE) program in an urban area in a state known for its production of eggs and poultry. As you plan an animal sciences unit related to poultry production, you begin to outline a field trip to a farm where chickens are raised. This facility hatches chicks and raises them to a certain size before they are sent to a different location to finish growing.

You soon realize that the field trip may not be possible. Your state has strict biosecurity guidelines, or procedures required to prevent diseases from being introduced to and spread within flocks, for poultry producers, including protocols that require visitors to shower before and after they visit a farm. In addition, you realize that your program is located several hours from the farm you want to visit, so you would have to arrange for learners to miss most of a day of instruction.

As you ponder what to do, you consider creating some type of technology-enhanced lesson. You know that agricultural organizations in your state have partnered with producers to plan such experiences. Learners in your program are given tablets to use for instruction and assignments, and your space is equipped with a classroom webcam and a projector.

You find the contact information for the farm's manager and begin to brainstorm.

The above scenario sets the scene for planning a virtual experience for the learners in your program. Using the information provided, create a lesson plan and a checklist (refer back to the *Example Checklist for Creating Virtual Tours*) for the experience.

As you plan, answer the following questions:

1. What platform and format will you choose? How will your lesson plan shape or be shaped by the delivery method?
2. What opportunities does the virtual experience offer that learners may not get from an in-person visit?
3. What potential barriers (technological, logistic, or pedagogical) may you encounter as you implement a virtual experience?

Reflective Questions

1. What teaching techniques might lend themselves to teaching agriscience? Which ones might be less applicable?
2. What technology do you expect to have access to in your future learning environment?
3. What resources would help you more effectively integrate technology into your lessons/curriculum?
4. How might your use of technology adapt to different groups of learners?

Glossary of Terms

- **TPACK:** Technological Pedagogical Content Knowledge; a framework used in education to describe the knowledge that teachers need to effectively integrate technology into their teaching practices which combines knowledge of subject matter, pedagogy, and technology to create effective and meaningful learning experiences for students
- **digital divide:** The gap between individuals, communities, and countries in terms of access to and use of information and communication technologies (ICTs); can be influenced by factors such as socioeconomic d, geographic location, age, and gender
- **accessibility:** The design and delivery of content, products, and services in a way that ensures they can be accessed and used by people with a wide range of abilities and disabilities; includes designing websites, software, and other technologies that are compatible with assistive technologies, such as screen readers, and creating content that is easy to understand for people with cognitive or visual impairments
- **assistive technology:** (AT); any device or software that helps people with disabilities or impairments to perform tasks that would otherwise be difficult or impossible; can include anything from simple aids such as magnifying glasses to complex communication devices and prosthetics
- **learning management system:** (LMS); a software application that provides a platform for managing and delivering online learning materials and courses; typically offer features such as course creation, assessment and grading tools, and communication tools to support online learning
- **hybrid instruction:** A teaching model that combines both in-person and online instruction; students may attend some classes in person and others online, or a single class may have some students attending in person while others attend remotely

- **synchronous instruction:** Teaching and learning that occurs in real-time, with both the instructor and students participating at the same time; can include activities such as video conferencing, live streaming, or chat rooms
- **asynchronous instruction:** Teaching and learning that occurs at different times, with students working on their own schedules rather than in real-time; can include activities such as recorded lectures, discussion, or online assignments
- **virtual field trip:** A digital experience that allows students to explore a location or environment remotely, using technology such as 360-degree videos or virtual reality (VR) simulations can provide students with immersive and engaging learning experiences that may not be possible to achieve in person
- **immersive technology:** Any technology that creates a simulated or augmented environment that surrounds the user, creating a sense of presence and immersion; can include technologies such as virtual reality, augmented reality, and mixed reality
- **virtual reality:** (VR); A technology that creates a completely immersive and interactive digital environment, often using a headset or other equipment to create a sense of presence and immersion; can be used for a wide range of applications, including entertainment, education, and training
- **augmented reality:** (AR); a technology that overlays digital information or graphics onto the real world, often using a mobile device or other display technology; can be used for a wide range of applications, including education, marketing, and entertainment

Figure Descriptions

Figure 8.1: Three circle Venn diagram. Circle 1 represents technological knowledge (TK). Circle 2 represents content knowledge (CK). Circle 3 represents pedagogical knowledge (PK). Between TK and PK is technological pedagogical knowledge (TPK). Between TK and CK is technological content knowledge (TCK). Between PK and CK is pedagogical content knowledge (PCK). All three circles meet in the center to create Technological pedagogical content knowledge (TPACK).

Figure 8.2: Three images. Tethered VR headset has a cord connecting an electronic headpiece that covers the person's eyes to other hardware. Standalone VR headset looks identical to the tethered VR headset, except it is cordless. Mobile VR headset is a headpiece that is made out of cardboard, covers the eyes, and has a place in front of the eyes to put a mobile phone.

Figure References

Figure 8.1: Technological pedagogical content knowledge (TPACK). Kindred Grey. 2023. Adapted and reproduced by permission of the publisher, © 2012 by tpack.org.

Figure 8.2: Tethered, standalone, and mobile VR headsets. Kindred Grey. 2023. CC BY 4.0. Includes Image by Karolina Grabowska from Pexels (https://www.pexels.com/photo/a-woman-wearing-a-virtual-reality-headset-while-holding-controllers-5207607/; Pexels license; 2020), image by fauxels from Pexels (https://www.pexels.com/photo/man-using-vr-goggles-3183187/; Pexels license; 2019), and image by othree from Flickr (https://flic.kr/p/o83BwL; CC BY 2.0; 2014).

References

Edutechnica (2020, February 2). LMS Data – Spring 2020 updates. *Edutechnica*. https://edutechnica.com/2020/02/02/lms-data-spring-2020-updates/

Fenton, M. (2017, October). Gamification and digital technologies for effective mathematics teaching. Paper presented at uLearn17 Conference, Hamilton, New Zealand.

Koehler, M. (2012, September 24). *TPACK explained*. TPACK.org. https://matt-koehler.com/tpack2/tpack-explained/

Koehler, M., & Mishra, P. (2009). What is technological pedagogical content knowledge (TPACK)?. *Contemporary Issues in Technology and Teacher Education*, *9*(1), 60–70. https://citejournal.org/volume-9/issue-1-09/general/what-is-technological-pedagogicalcontent-knowledge

Ninoriya, S., Chawan, P., & Meshram, B. (2011). CMS, LMS and LCMS For eLearning. *International Journal of Computer Science*, *8*(2), 1694-814.

Singer, N. (2017, March 2). Apple's devices lose luster in American classrooms. *New York Times*.

Whiteside, A. L., Garrett Dikkers, A., & Lewis, S. (2017). Overcoming isolation online: Strategies to enhance social presence in practice. In A. L. Whiteside, A. Garrett Dikkers, & K. Swan (Eds.) *Social presence in online learning: Multiple perspectives on research and practice* (pp. 180-187). Stylus, LLC.

9. ASSESSING AGRICULTURAL EDUCATION

Tiffany Drape

Setting the Stage

The end of the unit is nearly here, and we must provide progress reporting. A stack of papers sits on our desk, more documents are on our computer, and the wall is lined with projects we've had our learners working on. We think, "Don't forget all of the SAE projects that are off school grounds and CDE contest results." Immediate feelings of dread and being overwhelmed, not knowing where to start, and wondering how to even grade all of this overcome us. "Forget it, everyone gets an A," we think. The easy way seems like the path of least resistance so we can prepare for the next day and leave school at a reasonable hour. But we know our building principal will be observing us and asking us to provide examples of learner work, lesson plans, and how they were assessed so they can track our progress and report to the School Board and accrediting bodies. We go down the hall and ask another teacher what they do and how they grade. They email some rubrics. We scour the internet for assessments that are like the content we've been teaching hoping to find some fast help.

I can vividly recall this scenario my first year of teaching. No one taught me how to manage all the responsibilities I'd be tasked with, let alone how to then manage assessment, progress reports, and Individualized Educational Plan (IEP) accommodations, and then do a good job of reporting it to provide evidence of scholarly achievement. It was a watershed moment for me, and it will be for you too. Everything is new and unfamiliar, everything feels immediate, but much of it won't lead to providing evidence of assessment.

Objectives

The purpose of this chapter is to introduce our readers to assessing agricultural education. More specifically, the objectives are to:

- Briefly review *testing and assessment* in the United States.
- Explore the use of *formative assessment* as a tool to evaluate agricultural education to support its importance in secondary education.
- Investigate *backward design* as one way to provide evidence as a form of assessment.

Introduction

Understanding that every program is different and agricultural education is not a static program across states, regions, territories, or countries, this chapter aims to help us understand what assessment is, what counts as evidence, who determines this, and how we can use these to provide evidence of learning and career skill development for twenty-first century agricultural education. What I propose is formative, longer-term assessment, not the once-per-year test that our education system is currently built on. To provide evidence to our stakeholders, community members, and other folks who will want to know how we're training learners, formative assessment can drive home the importance and perhaps even necessity of agriculture education and Career and Technical Education (CTE) as an integral part of public education.

Overview of Assessing Agricultural Education

This overview provides a brief history of education standards at the state and national levels. Standards Based Education (SBE) rose to become mainstream in the 1970s and 1980s. Before this, most jobs in the United States required only an eighth-grade proficiency. With the rise of international trade and outsourcing of labor and manufacturing, the United States soon realized that it needed employees with higher-level skills and an eighth-grade education was no longer sufficient for these jobs.

Governors took initiative with drafting and implementing standards for their respective states and territories. By 1989, the National Council of Teachers of Mathematics published "Curriculum and Evaluation Standards for School Mathematics." In 1990, President Bush announced a set of education goals for the nation and the governors followed suit, forming a task force to monitor each state's goals. As the 1990s progressed, all but one of the states and territories developed their own unique statewide standards, and most either bought off-the-shelf tests for statewide use or created their own. Most states and territories worked with commercial testing companies to produce custom tests, frequently of much the same design as the most popular off-the-shelf tests. The tests satisfied the need of the accountability wing of the standards movement for an instrument that could be used to hold teachers accountable. In doing so, the tests fell far short of the kind of assessment tool that would provide incentives that fostered problem solving and critical thinking in the schools.

Note that standards vary on tribal, Indigenous, and other sovereign lands. The Bureau of Indian Education works as part of the US government to provide direction and management of all education functions, including the formation of policies and procedures, the supervision of all program activities, and the approval of the expenditure of funds appropriated for education on Indigenous lands. This encompasses 183 schools among 64 reservations in 23 states and 26 tribal colleges and universities.

As the twentieth century ended, the accountability movement came to dominate the standards movement in most states. States began building out comparison charts showing how schools compared to one another and to the state standards on the state tests. School districts everywhere began to find themselves under pressure to improve student performance based on the tests. This pressure was, of course, passed down to the schools and teachers. Many of the state tests were narrowly focused on facts and skills, rather than on a real understanding of the subject, critical thinking, and problem solving. Because of this, teachers focused almost wholly on test preparation and narrowed the curriculum. This depressed the achievement of many learners who would have achieved at higher levels if there had been no accountability system. Many teachers felt that the new tests and the accountability system that went with them were destroying effective teaching.

How did this narrow the curriculum?

The National Research Council (2011) concluded that the emphasis on testing yielded little learning progress but caused significant harm to both learners and teachers. Teachers who succumbed to the pressure were forced to only cover what would appear on the test, teach in less inclusive ways, and conform their teaching methods to the multiple-choice format that many tests are designed for, and many were driven out of the profession.

Narrowing the curriculum undermined student engagement, discouraged inclusive education, and harmed learners from low-income and underrepresented minority (URM) backgrounds, English language learners, and learners with disabilities. Children from white, middle- and upper-income backgrounds were more likely to be placed in "gifted and talented" or college preparatory programs where they are challenged to read, explore, investigate, think and progress rapidly while their less privileged peers often lagged even farther behind.

Missing from this standards-based model, but part of other educational assessment and curriculum delivery models, were: (1) high standards that incorporate a "thinking curriculum," like critical thinking; (2) assessments that teachers would like to teach to; (3) authority for the school principals and others who bear the burden of accountability to enact changes when necessary; (4) clear curriculum frameworks that would make it possible to build fully aligned instructional systems; and (5) investments in the tools and training the people would need to do the job.

In 2002, No Child Left Behind (NCLB) expanded, booming the high-stakes testing industry and becoming one of the major drivers in school performance determination. The impact of this law proved to be wide reaching, affecting every single school in some way. One of the major aspects of NCLB was the provision that required states adopt a system of accountability whereby learners, teachers, administrators, and schools are evaluated annually based on learners' standardized test performance and that consequences follow when student scores are low or annual gains in school achievement are not made.

Negative consequences include narrowing the curriculum, teaching to the test, pushing learners out of school, driving teachers out of the profession, and undermining student engagement and school climate.

Learners from low-income and minority-group backgrounds, English language learners, and learners with disabilities, are more likely to be denied diplomas, retained in grade, placed in a lower track, or unnecessarily put in remedial education programs. They are more likely to receive a "dumbed-down" curriculum, based heavily on rote drill and test practice. This ensures they will fall further and further behind their peers. Many drop out, some ending up in the "school-to-prison pipeline." On the other hand, children from white, middle- and upper-income backgrounds are more likely to be placed in "gifted and talented" or college preparatory programs where they are challenged to read, explore, investigate, think, and progress rapidly.

As a result, teachers and administrators feel enormous pressure to ensure that test scores consistently rise. Schools narrow and manipulate the curriculum to match the test, while teachers tend to cover only what is likely to be on the next exam. Methods of teaching conform to the multiple-choice format. Education increasingly resembles test prep. It is easy to see why this could happen in low-scoring districts. But some high-scoring schools and districts, striving to keep their top rank, also succumb. The pressure is so great that a growing number of administrators and teachers have engaged in various kinds of cheating to boost scores.

High-Stakes Testing

We can't talk about learning standards without talking about high-stakes testing. Since its inception, it's grown into a billion-dollar business. SAT, ACT, LSAT, GRE, and MCAT are a few of the acronyms we might be familiar with and perhaps have even had to take yourself. High-stakes testing has arguments both for and against them and the debates about costs, equity, and access continue. Whether we like them or not, our learners will have to take them or have taken them already. Understanding their role in public education and how we can use them as part of our agriculture programming can be helpful in providing evidence to our stakeholders in our school district.

These tests are not only used to assess student aptitude and achievements, but also to inform how funding is funneled, assess curriculum and instruction, and help make predictions about how successful a student will be after they graduate. Most folks never take the time to investigate how these tests came to be adopted en masse or if they are in fact valid and reliable tools at determining aptitude and achievement of learners or a school. We know that high-stakes testing is not the panacea but the education system we may work in will use these tests as input to make decisions each year.

We know that high-stakes testing affects things like a student passing to the next grade level and choosing a school or moving within a district and schools being able to receive funding for special services and IEP accommodations. The list is endless. In some schools, learners will take as many as 113 standardized high-stakes test from K-12.

Most Americans trust that these tests were crafted with the utmost care and accurately report on student achievement in a reliable and valid way. For the most part, these tests have been integrated into the current education system. Most citizens don't question the information generated from these tests and how they are used.

Opportunity for Agricultural Education

This is the opportunity for agricultural education to lead on the reimagining of what the future of standards might look like. Compulsory education is one of the longest running experiments in our country and agricultural education should continue to be part of this basic education that school-aged children deserve. Creating compelling, reliable methods to formatively assess learners is one way to ensure that the importance of agricultural education does not get lost. We can teach learners how to critically think, problem solve, and provide long-term evidence of learning. Agricultural education can adapt with these changes. Many agricultural education teachers are already dual certified, offer college level courses to empower learners with inclusive education, and can be asked to teach one or more classes outside of their certification area, depending on their state/territory certification policies. With the wide variety of agricultural education classes, dual enrollment, college credit, and the integration of Future Farmers of America (FFA) and Supervised Agricultural Experience (SAE), agricultural education has expanded the narrow curriculum in a manageable and actionable way. Now the field needs to prove why it works and model how it can work while working with and respecting NCLB, SBE, and high-stakes testing.

Agricultural education and CTE offer valuable skills, but they can be hard to justify without evidence. Not all administrators view evidence the same way. A ribbon with a county fair SAE project or win at a Leadership Development Event (LDE) is not always considered assessment. As budgets shrink, test scores continue to drive funding, and resources continue to be stretched, agricultural education must take advantage of the work it's currently doing to create the program we envision. Whether we're a veteran teacher in that school or it's our first job, the goal of assessment should be to leave it better than we found it. What "better" means is up for interpretation.

In every education setting we'll work in, both formal and informal, there will always be the trend or the popular thing at the time. Word walls, learning objectives, book clubs, and a host of other district mandated teaching strategies will be employed by administrators, and we'll be expected to conform to these things. How can we design assessment to endure long past the newest trend? How do we incorporate the new trend without feeling like we need to start from scratch?

We're Not Grading, We're Collecting Evidence

Let's reframe our thinking first. We're not grading assignments. We're collecting evidence to determine student progress and pointing them toward their next steps. The mental switch alone will provide us the time, space, and permission to figure out if we want to prioritize assessment. When we think of it as providing feedback and pointing in the next direction, it no longer turns into this "thing we have to do" but will hopefully make it the thing we need to inform the learners, ourselves, and other interested stakeholders. We don't have to collect all the evidence by ourselves either. Expanding the relationship with our learners to engage them in their own progress through keeping records, actively reflecting, exhibiting incremental progress, and sharing those things with us will help take some of the burden from us. Providing feedback is not a one-way conversation, it should be reciprocal.

What Counts as Evidence?

Almost anything can count as evidence. Let's start with existing evidence. New teachers rarely look at what already exists regarding assessment and with learners who may have taken prior coursework in that program. Other existing data can be student records, IEP plans, and prior results from FFA and SAE work. This can serve as baseline evidence and gives us and the learners a place to begin identifying what milestones or progress they can work toward.

Some say that any assessment framework and any assessment model that tracks progress against individual learning intentions and targets counts as evidence. This operates under the assumption that these models and frameworks exist, which they may not be the case during our first year or three of teaching. Evidence comes from two principal sources: direct observation and the examination of artifacts.

If we need a nonexhaustive list of what could count as evidence:

- Images, still and moving
- Audio evidence
- Interviews, in audio or transcribed form
- Group discussions, in audio or transcribed form (check school policies on recording)
- Documents
- Observations
- Field notes
- Data from think-aloud protocols
- Anecdotes
- Answers to questionnaires
- Record books
- Results from CDE's or other leadership events

We also need to consider what counts as evidence for teacher evaluations. The two are intimately linked. We may want to inquire about this with our mentors, colleagues, and building principal to help us understand the evaluation itself and what's being evaluated. The task then becomes figuring out what to do with it, how to assess it, and then how to package and communicate it to the various stakeholder groups.

Formative and Summative Assessment

While there are many kinds of evidence from the list above, how we organize and report evidence falls in one of two camps: formative or summative assessment. We can put a play on words to help us remember the difference between formative and summative. Summative assessment refers to a summary at the end. Summative assessment, which is what high-stakes testing is, is the one time, one test that determines a multitude of aspects of a student, their teachers, their school, and their performance across that class or grade.

Formative refers to assessment that is formed along the way or incrementally. Instead of one test to determine the success or failure of a student, I'm proposing we take a formative approach to assessment.

Here's why: Agricultural education is an evolving industry, and an evolving curriculum goes with it. The curriculum is always adapting, improving, learning new things that add to our collective education. As the agriculture industry changes, we must adapt as well. Formative assessment gives us the autonomy to adapt without losing the rigor and evidence-based collection.

What would happen if we could collect that evidence incrementally instead of waiting until the end of the marking period or term and then panicking because grades are due? As we look back at the pile of "work" or evidence, many teachers look at it and say or think "What did I even do with these kids the past X number of weeks?" Some are more organized, but new or young teachers are often so overwhelmed by everything that they forget that the evidence they provide informs decisions about their learners, their own employment, the school, and the community. There are so many aspects to be mindful of when starting out, assessment often falls by the wayside until it doesn't, and the principal or team leader needs that evidence for reporting.

We may be asking ourselves several questions: How do we do this? Collect intentional evidence to provide formative evidence?

Let's think about it backward. At the end, what do we want our learners to know, or what skills should they have acquired, and what evidence do we want to be able to provide to prove incremental progress?

Learning Confirmation

Let's use this sample objective to help guide us:

- **Goal for the learner:** Learners will be more proficient public speakers.
- **Objectives:** Understand, describe, and report on agricultural issues.
- **Task:** Apply information to design and develop a prepared public speech for a contest.

Backward Design: A Primer

What if we designed our formative assessments using a backward design? Thinking about what we want the student to understand, explain, and expand at the end of that course can help us work from the end back to day one (figure 9.1). Along the way, we can collect the incremental pieces that will account for the evidence we'll use in the formative assessment of the learners. Since evidence is broadly defined, we can move away from that "traditional test" we've been taught and expand what we might want to collect that counts as evidence. Instead of one recording of the student presenting their speech at the end of the course, recording at several points during the class will create a trail of evidence that we can use to assess student progress and with stakeholders to show incremental progress over time. We may find that sometimes those "traditional tests" are still useful, but they're not the only or best way to provide evidence of student progress toward that objective at the end of the course. What other pieces of evidence could we collect to provide concrete progress in our learners throughout the course?

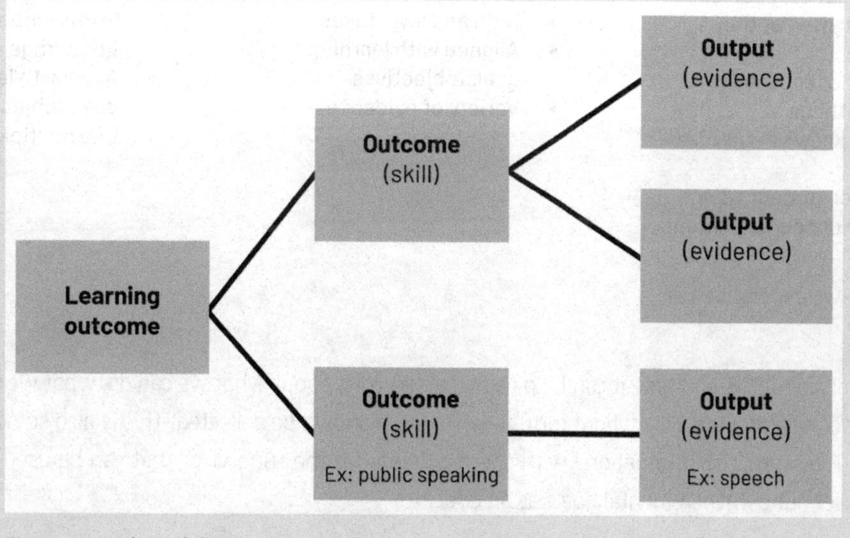

Figure 9.1: Backward design.

We can start planning with the final outcome in mind to guide the class, account for changes along the way, and offer places for us as the teacher to make adjustments. We can ask ourselves several questions to help frame our thinking. Start with the "need to know" questions first: What do learners need to understand at the end of the unit or course? What skills will they acquire through completing the unit or course? At what points will feedback be important to give so learners can keep moving forward? How are we ensuring this is inclusive and will accommodate learners who may need more time to grasp concepts or have other needs?

Move onto what might be "nice to know" questions: What might be nice to include if we have space? What can be removed if something happens that hinders the progress?

Finally, "what do we need" questions: What do we need to learn to teach our learners? What resources do we need to teach the content? Where can we find the information and resources we'll need?

Laying the answers out on note cards or organizing using a visual aid can help refine and prioritize what we might teach. As an example, if lessons are taking longer than anticipated, we can go back to our design and decide which activities and pieces of evidence will provide the information we need and what can be changed, edited, or cut. It will not change the goals or objectives of the class; it will only adjust and adapt the formative evidence we collect along the way. (See figure 9.2.)

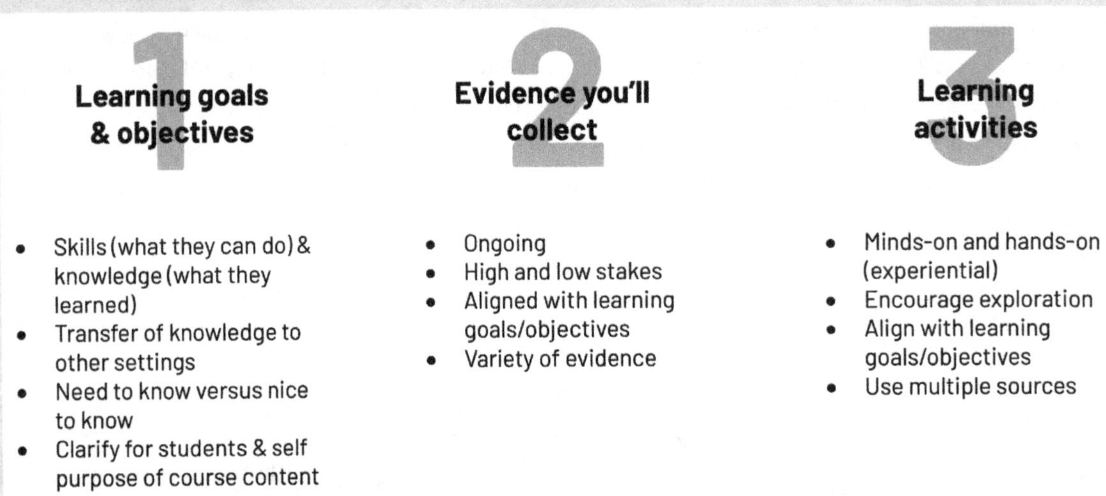

Figure 9.2: Backward course design.

When we think about the end first, it can help us make decisions about what we can do, what we have time for, and identify points along the way where formative evidence should be collected. This is also something we can provide for our own teacher evaluations with specific points for the supervisor that can be used to gauge our progress and set our own incremental goals as an educator.

Applying the Content

Questions to ask:

1. What do we want learners to understand at the end of the unit/course? What evidence do we need to exhibit that proficiency or skill?
2. Next, what steps or skills do learners need to learn to meet objective one?
3. What pieces of evidence from class activities do we need to collect to provide incremental progress to meet objective one?

Deliberately asking ourselves these questions can help us distill what's important for a student to be competent in versus what's less important. It can then inform our decision-making process on what we'll count as evidence, at what point(s) it will be collected, and how it will be assessed. In short:

1. Identify the desired results (big ideas and skills).
2. Determine acceptable levels of evidence that support that the desired results have occurred (culminating assessment tasks).
3. Design activities that will make desired results happen (learning events).

Secondary questions after we've made this initial list or document could include how much time these activities take, what will count as evidence, what supplies/materials we need, and what considerations exist for students with IEP/special needs.

Assessing the Learning Activities

Let's be honest, there is no right or perfect way. Looking for a panacea will not be found. What we're sharing are some basic guiding tenants to begin thinking about, practicing, and implementing to help us begin our teaching career. Our teaching style, the community we work in, and the learners we serve are a few components that will inform how we assess learners. First know that we can (and should consider) using existing rubrics to assess learner work. Returning to the example of a public speaking course, the MANRRS Public Speaking Contest, National FFA Public Speaking Contest, or any related guidelines can be used and adapted to fit our student's needs. We don't need to reinvent the wheel and we can make changes to adapt preexisting models to our learners and their work. We can use the same rubric over the life of the public speaking class or unit and document student progress over time. This will show areas of improvement for both us and the learner and provide a long-term evidence trail over the life of the class or unit.

Formative assessment not only means collecting incremental evidence as a process over time but giving specific feedback to learners that is then turned into student action, assessing regularly, sometimes even daily, and taking the data to modify the learning plan and instruction for the learners. Rapid feedback loops must be incorporated to offer learners regular feedback. Feedback can be given in multiple ways and doesn't have to be long. It could be as simple as figuring out how to get a learner to stop using filler words (e.g., um, like) and can be documented through recordings that the student then reviews and works to alleviate their use. Having those recordings will provide incremental evidence that the student did improve their public speaking skills.

Formative assessment allows for the student to participate as well. When we look for progress over time, the learner can actively participate in deciding what they deem as progress. Using fewer filler words, being able to reflect on class activities, and making adjustments should involve the learners so they can not only participate in their own learning, but also participate in their progress to achieve that learning.

- What would our backward design look like using the example of the public speaking course?
- What changes or edits are needed?
- What pieces of evidence would we collect to show as evidence?

Using Formative Assessment for Teacher Evaluations

We'll collect a lot of small pieces of incremental evidence from our learners. We may have audio recordings of speeches, drafts of speeches with edits, and perhaps even a result from the CDE contest with a picture of that learner presenting in the contest. Our learners can help with this as well, providing drafts, reflective feedback, and organizing any of the evidence. All these together can tell a compelling story that can be used to evaluate our progress as a teacher.

It's one thing to be able to report that we had a successful student at the contest, but if our class had more than one student, we should consider including their evidence since it likely showed improvement from day one to the end of the class. That collection of audio files, speech drafts, and self-reported feedback from the learners puts together a compelling set of evidence that can speak to our effectiveness as an educator and provide data for our evaluations (see figure 9.3).

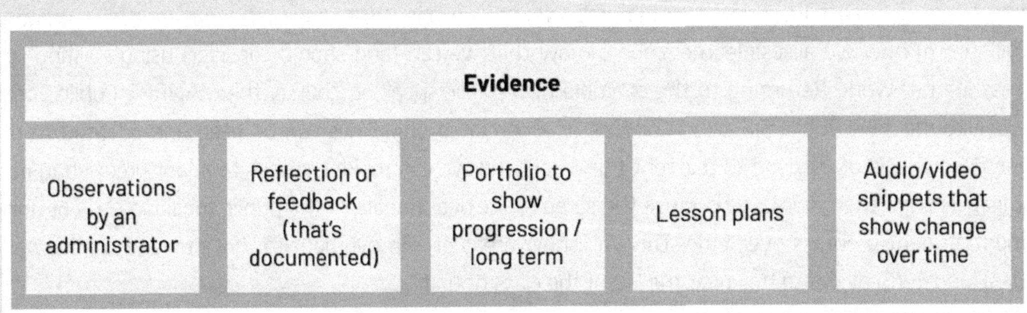

Figure 9.3: What counts as evidence for teacher evaluations.

Peer review/exchange, a type of assessment: Take this existing curriculum and work through it. Does it meet the learning objectives set forth? Please provide at least two pieces of evidence in the curriculum that meet those objectives and identify them. Exchange with your peer(s) and work through their work. What agreement did you have? Where did you find different evidence to support the learning objectives? In what ways is curriculum subjective? When we interpret, what should we be mindful of when working with learners? What should we be mindful of when working with administrators and other stakeholders?

Reflective Questions

In Summation and Reflection

We can only handle so much during our first year of teaching. Instead of trying to "do it all," decide on one or two places to begin evaluating and amending the assessment plan. It will take practice and multiple iterations. We'll learn as much, if not more, than our learners.

If we go back to our introduction of the teacher with piles of work but no way to assess, what might they do to triage the situation? Since this could be us soon, know that it is common, it takes practice, and it's OK to ask for help and guidance.

What do we want agricultural education to look like in the future? What steps might we need to undertake related to assessment to get to that vision shared?

Glossary of Terms

- **formative:** Monitor student progress during the program or intervention
- **summative:** Evaluate student learning at the end of the program or intervention

Figure Descriptions

Figure 9.1: Flow chart. Learning outcome maps to an outcome or skill, such as public speaking. That outcome maps to an output ot evidence, such as a speech. Each learning outcome can have multiple outcomes. Each Outcome can have multiple outputs.

Figure 9.2: Start with learning goals and objectives. This includes: (1) skills (what they can do) and knowledge (what they learned), (2) transfer of knowledge to other settings, (3) need to know versus nice to know, and (4) clarify for students and self purpose of course content. Then think about evidence you'll collect. This includes: (1) ongoing, (2) high and low stakes, (3) aligned with learning goals/objectives, and (4) variety of evidence. Lastly, learning activities. This includes: (1) minds-on and hands-on (experiential), (2) encourage exploration, (3) align with learning goals/objectives, and (4) use multiple sources.

Figure 9.3: Evidence includes: observations by an administrator, reflection or feedback that's documented, portfolio to show progression or long term, lesson plans, and audio or video snippets that show change over time.

Figure References

Figure 9.1: Backwards design. Kindred Grey. 2023. CC BY 4.0.

Figure 9.2: Backward course design. Kindred Grey. 2023. CC BY 4.0.

Figure 9.3: What counts as evidence for teacher evaluations. Kindred Grey. 2023. CC BY 4.0.

References

Danielson, C. (2007). *Enhancing professional practice: A framework for teaching.* ASCD.

Evidence for Learning. (2020, December 3). *Evidence for Learning in action!* https://www.evidenceforlearning.net/eflinaction/

The National Research Council. (2011). *Incentives and test-based accountability in education.* The National Academies Press.

National Academies of Sciences, Engineering, and Medicine. 2011. Incentives and Test-Based Accountability in Education. Washington, DC: The National Academies Press. https://doi.org/10.17226/12521

Nichols, S. L., & Berliner, D. C. (2007). *Collateral damage: How high-stakes testing corrupts America's schools.* Harvard Education Press.

PBS. (2020, December 15). *Americans Instrumental in Establishing Standardized Tests.* https://www.pbs.org/wgbh/pages/frontline/shows/sats/where/three.html

Stateuniversity.com. (2021, March 23). *Standards Movement in American Education.* https://education.stateuniversity.com/pages/2445/Standards-Movement-in-American-Education.html

10. APPLIED LEADERSHIP DEVELOPMENT THROUGH FFA

Laura L. Greenhaw and Paula Faulkner

Setting the Stage

Sherry and Jose teach agriculture in a high school with a student population of approximately six hundred. Nearly half of the students in the school are enrolled in at least one agricultural education course. Sherry recently submitted a purchase order for program affiliation FFA membership so all students enrolled in an agricultural education course will be members of the FFA. Although their FFA chapter has utilized program affiliation membership previously, the new principal questions the purchase order, stating that students in other clubs must pay their own dues to participate in extracurricular activities. Sherry and Jose schedule a meeting with their new administrator to explain how FFA is an intracurricular component of the holistic agricultural education program. They prepare several examples that demonstrate FFA is fundamental for teaching and learning in agricultural education.

Objectives

Upon completion of this chapter, learners should be able to:

- Explain FFA as an intracurricular component of the complete agricultural education model.
- Describe the structure and components of FFA.
- Identify competencies taught through FFA activities.
- Illustrate efforts and opportunities for diversity and inclusion in FFA.
- Utilize FFA programs and activities as experiential learning opportunities for learners.

Introduction

FFA is a national organization that prepares members for premier leadership, personal growth, and career success through school-based agricultural education. The official name is the National FFA Organization. The letters "FFA" stand for Future Farmers of America, an important part of our history and heritage.

FFA develops youth with career and leadership development and helps them discover their talent through hands-on, minds-on experiences, giving them the tools for future success.

Former FFA members are teachers, scientists, veterinarians, government leaders, entrepreneurs, financial professionals, international business leaders, and professionals in many fields (National FFA Organization, 2021).

Overview

In 1988, the National Academy of Sciences Committee on Agricultural Education in Secondary Schools published *Understanding Agriculture: New Directions for Education*. Authors described secondary vocational agriculture programs as consisting of three parts, including "classroom and laboratory instruction, supervised occupational experiences (SOEs), and membership in the National FFA" (p. 2). More than three decades later, we continue to reference these three components, albeit with slightly modified language. Today, school-based agricultural education (SBAE) is grounded in a three-circle model that includes the classroom and laboratory, Supervised Agricultural Experience (SAE), and FFA (figure 10.1). FFA, the student organization, is a unique component of the three-circle model designed to teach and develop learners' "premier leadership, personal growth, and career success" (The National FFA Organization, 2021). Through a variety of events, competitions, awards programs, conferences, degrees, and other methods, the FFA infuses value in all learners' education. When combined with classroom and laboratory instruction and SAE projects, FFA connects, reinforces, and extends learning through simultaneous practice of leadership and technical agricultural knowledge, skills, and abilities.

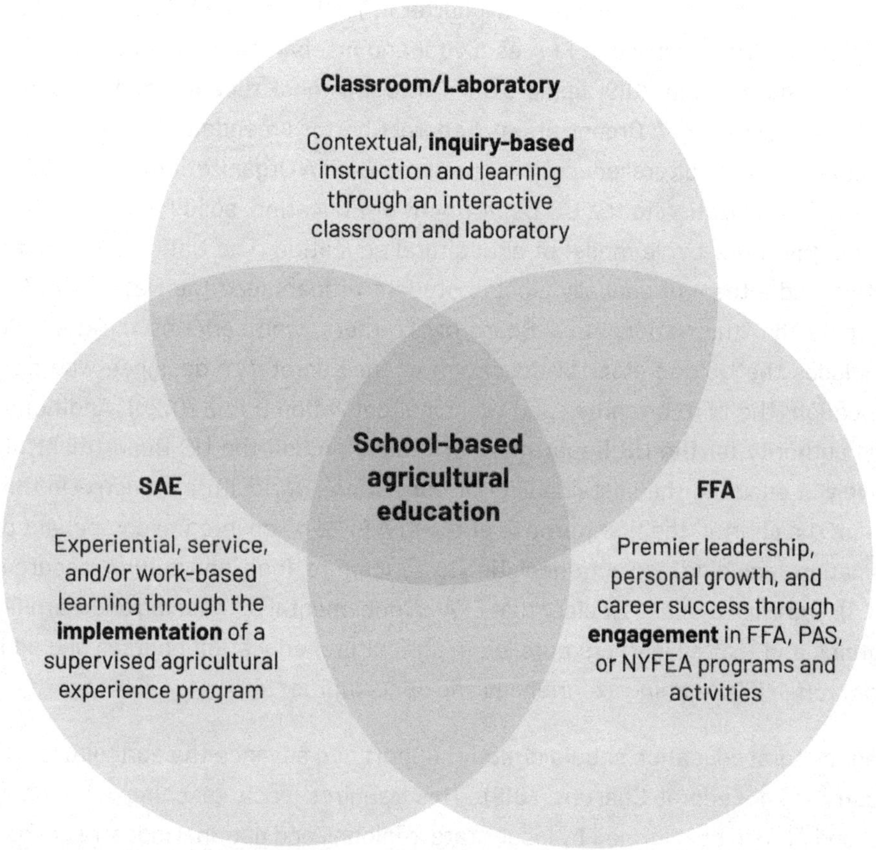

Figure 10.1: Three-circle model.

In this chapter, we will discuss how FFA is an intracurricular component of a comprehensive SBAE program, describe the structure and components of FFA, identify competencies taught through FFA, illustrate diversity and inclusion efforts and opportunities in FFA, and demonstrate how to utilize FFA as an effective teaching strategy.

FFA Is Intracurricular

Intracurricular means "within the curriculum." Unlike many other school organizations students can engage in outside of class, FFA is *not* an extra component of the educational experience. Rather, it is an equal contributor to the whole agricultural education program, integrated with classroom and laboratory instruction and supervised agricultural experience (SAE). Why is FFA different from extracurricular programs? One important difference is that the federal government recognizes the National FFA Organization as a crucial component of SBAE.

The National FFA Organization received a federal charter in 1950 when the US Congress passed Public Law 81-740 (FFA, 2019a). The law established FFA as a vital and inseparable component of SBAE. The charter was revised in 1998 and again in 2019, being signed into law most recently by President Donald Trump. Public Law 116-7, the "National FFA Organization's Federal Charter Amendments Act," reiterates that FFA is an integral component of agricultural education. The National FFA Organization's federal charter clearly and inextricably links the organization to the US Department of Education, solidifying its critical role as one of the components in the three-circle model of agricultural education. The National FFA Board of Directors, National FFA staff, and a team of annually elected student officers lead the National FFA Organization. It is important to note that the National FFA Board of Directors, who serve as the governing body of the organization, includes the US secretary of education or the secretary's designee who has experience in agricultural education, the FFA, or career and technical education (FFA, 2019b). Additionally, the charter provides federal authority for the US Department of Education and the US Department of Agriculture to work cooperatively in ensuring the sustainability of our nation's agriculture industry. In the most recently passed iteration of the charter, the first purpose of the FFA is "to be an integral component of instruction in agricultural education, including instruction relating to agriculture, food, and natural resources" (FFA Federal Charter, 2019, p. 1). The charter makes it clear that FFA is fundamental to teaching and learning in agricultural education programs and that all learners enrolled in agricultural education courses should have access to and be engaged through FFA learning experiences and opportunities.

School-based agricultural education should directly support and advance the agriculture, food, and natural resources industry (FFA Federal Charter, 2019). This requires each teaching component (classroom/laboratory, SAE, and FFA) to be informed by local, state, regional, and national industry needs. Let's consider an example.

> Keisha teaches in a small public school in a rural area in the Southwest in the United States. Much of the local agriculture industry is traditional beef cattle production. The region also supports a significant amount of hunting, including several big game outfitters. Appropriately, Keisha has worked with her administration and school board to offer classes including science of large agriculture animals, agricultural economics and business management, environmental science and natural resources, science of food products and food processing, and science of wildlife and forestry management. She collaborates with local stakeholders to develop and support SAE opportunities for her students, such as ownership SAEs in beef and alfalfa production, placement SAEs at meat processing and taxidermy businesses, and research SAEs on controlled burns and wildlife management strategies.

Finally, Keisha incorporates FFA competitions and events into her curriculum. Students compete in several career and leadership development events, apply for proficiency awards, attend conferences and conventions, and develop an annual program of activities that reflects the culture of the area and the developmental interests of the students. How do we know that Keisha is using FFA to teach her students? Each of the FFA activities compliments and supports the classroom/laboratory instruction and SAE opportunities. Keisha's students compete in livestock evaluation, meats evaluation and technology, farm and agribusiness management, environmental and natural resources, agricultural sales, and agricultural technology and mechanical systems Career Development Events. They apply for proficiency awards corresponding to their SAEs, and several students conduct and present agriscience fair projects developed from assessing and addressing local industry needs. Keisha works hard to ensure students experience a comprehensive education in her program, through a combination of classroom/laboratory instruction, SAE, and FFA.

FFA Is Experiential Learning

John Dewey believed that knowledge was constructed through real-life experiences in the social environment. Teachers are responsible for facilitating learning experiences and organizing the content to be learned within those real-life experiences. The social environment learning occurs in is paramount, according to Dewey, and social interactions should be carefully planned to facilitate the development of relationships, particularly those between adults and youth (Roberts, 2003). Dewey also believed that learners should acquire knowledge through their own personal experiences rather than have it dictated to them from outside sources such as textbooks or experts. Knowledge gained from past experiences, then, is used to teach about the present. Equally important is the organization of content for the learner. Subjects should not be isolated but integrated so that the learner makes connections between the content and real life where knowledge and skills are used simultaneously. For example, in real life, interpersonal skills are used simultaneously with technical skills by Extension agents working with farmers and ranchers to improve their crop yield.

Like Dewey, Kolb (1984) defines learning as "the process whereby knowledge is created through the transformation of experience" (p. 41). According to Kolb's Experiential Learning Theory, learners progress through a four-part cycle to transform experience into knowledge. First, learners encounter new experiences or reinterpret past experiences. Then, they reflect on their experience, particularly how it may differ from their previous understanding. Abstract conceptualization follows, resulting in a new or modified idea or understanding. Finally, a learner actively experiments with their new idea or concept. (See figure 10.2.)

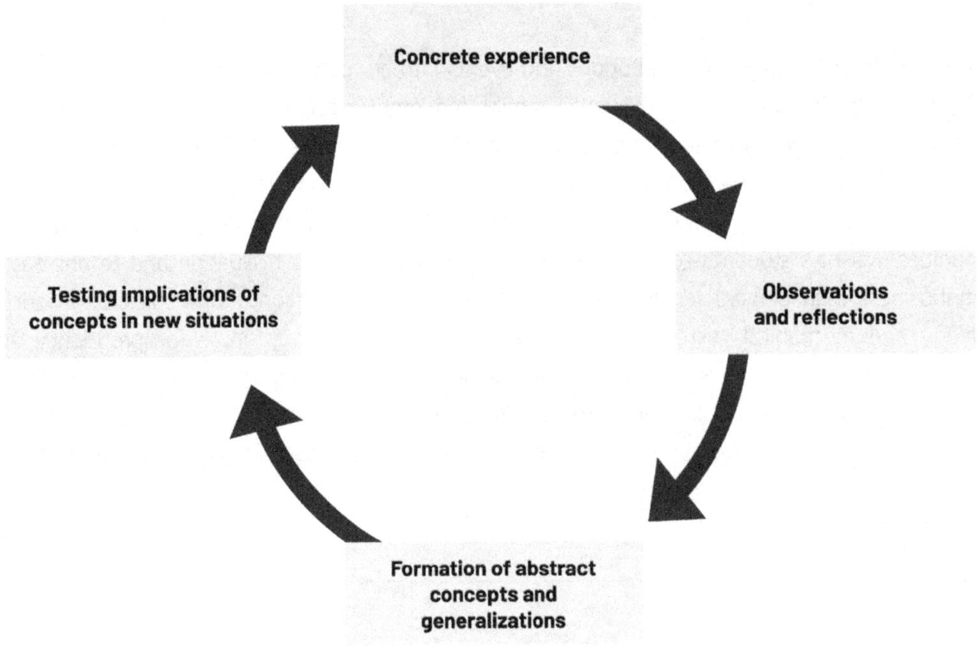

Figure 10.2: Kolb's experiential learning model diagram.

Experiential learning has been a foundational tenant of agricultural education since its inception (Baker et al., 2012). FFA provides a wide variety of experiential learning opportunities to engage learners at varying levels of readiness. The teacher should select experiences that situate learning in the context of local, state, regional, and national agriculture, food, and natural resources systems so learners can draw from their own experiences to build and expand knowledge and skills.

Moreover, as a student organization, FFA provides the social environment Dewey believed was so critical to learning. Although intracurricular, FFA activities often occur outside the classroom and can involve external stakeholders, thus fostering positive adult-youth relationships. Finally, these experiences demand learners use knowledge from multiple subjects at the same time, applying technical content coupled with leadership and interpersonal knowledge and skills.

Although FFA provides a well-built vehicle for learning in agricultural education, the burden of keeping the vehicle on the road to learning lies with the teacher. "Teachers must be present and mindful throughout the experiential learning process to guide and direct the learning process" (Baker et al., 2012, p. 7). The teacher is responsible for ensuring experiences are educative, because, as Dewey argued, not all experiences are (Roberts, 2003). The agricultural educator shoulders two responsibilities in Dewey's experiential learning model: knowing the subject matter and knowing the learner. Further, a teacher must skillfully guide the learner through each phase of Kolb's model (Baker et al., 2012). Therefore, planning is essential to creating worthwhile learning experiences. This requires the teacher to:

- Understand learners' current knowledge,
- Determine appropriate learning environments,
- Allow the correct amount of learner autonomy,
- Accurately estimate learner readiness such that educational experiences are not beyond the scope of learners' current ordinary life experiences,
- Organize learning experiences toward a specific end or outcome,
- Facilitate learners' full extraction of meaning from experiences so they can apply the knowledge to future situations, and
- Guide educational experiences in a way that fosters students' life-long learning capabilities (Roberts, 2003).

In the following sections, we will explore specific FFA learning experiences teachers can use to plan and guide learners through the experiential learning process. We begin with the structure of the FFA.

The FFA Organizational Structure

The National FFA Organization is structured on three levels (FFA, 2021). Learners can engage in activities and experiences through their local FFA chapter, their state association, and at the national level. The National FFA Organization staff work with a board of directors, six student national officers, national convention delegates, and agricultural education stakeholders to serve and support over 760,000 student members. All fifty states as well as Puerto Rico and the US Virgin Islands have state associations that charter local FFA chapters. As of 2020, nearly nine thousand local FFA charters had been granted.

While all three levels function under the National FFA Organization's constitution and bylaws, state associations are led by their annually elected student officers and advised by the state supervisor of agricultural education. State associations have the autonomy to create unique learning opportunities that best fit their state's needs, including competitions, awards, programs, and leadership structures within the bounds of the National constitution and bylaws. Likewise, local FFA chapters are led by annually elected chapter officers and are advised by the agriculture teacher. Each local chapter creates its own unique program of activities (POA) to serve and support its learners' developmental needs. The three-tiered organizational structure's flexibility supports local agricultural educators as they plan relevant educational experiences for their unique learners (Dewey, 1938; Roberts, 2003). To learn more about the National FFA Organization, its constitution and bylaws, and how local FFA chapters are chartered, please visit the National FFA Organization website https://www.ffa.org.

Competencies Taught through FFA

The National FFA Organization promotes premier leadership, personal growth, and career success. Learners acquire and apply technical knowledge combined with leadership and interpersonal knowledge and skills through FFA activities. Agricultural educators are responsible for organizing learning experiences relevant to the local, regional, state, and national industry needs, which provides learners with technical content. Technical content may be food science, urban agriculture, animal production, agricultural mechanics, or a multitude of other topics relevant to food, agriculture, and natural resources. Leadership and interpersonal competencies are integrated into the learning and practice.

Examples of leadership and interpersonal competencies include but are not limited to oral and written communication, critical thinking, problem solving, teamwork, personal responsibility, meeting facilitation, conflict management, recording keeping, innovation, creativity, goal setting, collaboration, relationship building, and others. Learners practice technical skills simultaneously with leadership and interpersonal skills just as they occur in real life. For example, students applying for a proficiency award in agriscience research not only demonstrate technical competence on their research topic, but they also demonstrate recording keeping, communication, and other leadership and interpersonal skills.

Students competing in the forestry career development event (CDE) must call upon their technical knowledge about forestry while simultaneously practicing teamwork, communication, time management, and other leadership and interpersonal competencies. Local chapter officers and members apply proper parliamentary procedure to conduct the work of their organization in meetings but also to practice collaboration, meeting facilitation, relationship building, and other leadership and interpersonal skills as the work is carried out in accordance with their program of activities.

FFA Learning Experiences

Membership and Degrees

The National FFA Organization provides a broad structure within which agriculture teachers can plan and organize learning experiences. All FFA experiences begin with learners enrolling in an agricultural education course and securing membership in the FFA organization. Most often, this requires the school to have a chartered local FFA chapter. However, some state associations and local chapters may allow students to join FFA chapters in neighboring school districts if their school does not have an FFA chapter. Although FFA is considered an integral and inseparable component of the agricultural education program, annual membership dues are still required. Membership dues are levied at the national level and may include additional charges from the state association and local chapter.

Currently, there are two ways to become a member. In some chapters, dues are paid for each individual student who wishes to join the organization. Students themselves may pay their dues or the chapter may utilize fundraising or other means of financial assistance to pay dues for students who wish to become FFA members. Other local chapters pay program affiliation fees that provide FFA membership for all learners enrolled in one or more agricultural education courses, based on a progressive fee structure. This allows all students enrolled in agricultural education courses to be FFA members, regardless of their ability or willingness to pay an individual annual membership fee. All membership dues and program affiliation fees are established by a majority vote of eligible voting delegates at each level of the organization: national voting delegates at the National FFA Convention, state voting delegates at the state convention, and eligible members at a regular local chapter meeting.

Membership provides the first and most basic teaching opportunity. FFA members are eligible to progress through five degrees of membership based on individual achievement (FFA, 2021). Each degree of membership advances from the previous as the learner achieves the minimum qualifications and applies for each degree. In order, membership degrees are Discovery, Greenhand, Chapter, State, and American. It should be noted that Discovery degrees are intended for state associations and local chapters that offer membership to middle school students. Teachers can use degrees of membership to organize educational content as learners acquire knowledge through experience. Since each degree builds upon achievements from the previous degree, the learning is naturally situated within the learner's realm of understanding and linked to each learner's unique previous experience (Dewey, 1938; Roberts, 2003). Moreover, learners pursue each degree's qualifications in a context of their choosing, further ensuring that knowledge is built from their personal experiences.

Let's consider another example.

> Sarah is enrolled in an agribusiness course at her high school. She is an active FFA member and, with her teacher's guidance, has developed an SAE. Last year, as a freshman, she earned the Greenhand degree by successfully meeting the National FFA constitution's requirements and submitting a written application to her local chapter. One of the requirements was to develop a satisfactory plan for an SAE. Sarah discussed her interests and experiences with her teacher and after conducting research with her teacher's guidance, she decided she was most interested in beekeeping. Her teacher introduced her to a local beekeeper, who offered Sarah an opportunity to work with him and learn more about his business.
>
> This year, Sarah plans to apply for her Chapter FFA degree. She enrolled in the agribusiness course to learn more about owning and operating her own business. In addition, she works with the local beekeeper a few hours a week, earning a little money and gaining a lot of experience. She attends the local farmer's market with the beekeeper, where she assists with teaching the public about honeybees.

She plans to use the money she earns to purchase her own hive and rent some space from the local beekeeper to keep her hive with his. With her teacher's help, she has laid out a four-year plan to advance through the degrees of FFA membership, eventually earning her American FFA degree.

Competitive Events and Awards

The National FFA Organization offers a robust program of awards and competitive events that can be implemented at all three levels (local, state, and national). Competitions and awards are available for individuals, teams, and chapters. Competitive events are generally separated into Career Development Events (CDE) and Leadership Development Events (LDE).

As previously stated, the National FFA constitution and bylaws allow for some autonomy at the state and local levels to develop and offer competitive events and awards that reflect both the industry's and learners' unique needs. A current list of national competitive events and awards can be found on the National FFA Organization website, www.ffa.org. Teachers should contact their state association for a current list of state competitive events and awards. Some examples of unique events offered by state associations include the Quartet contest in Alabama, citrus evaluation in Florida, tractor troubleshooting in Vermont, milk quality evaluation in North Carolina, meat evaluation in Pennsylvania, and cotton judging in Texas.

Competitive events provide incentive and require students to combine technical knowledge and skill with leadership and interpersonal knowledge and skill. Some competitions require a team, while others are for individuals. Competitive events expand learners' social environment as FFA members may receive coaching from community members who are industry experts and travel to various locations to compete and interact with FFA members and advisers from other chapters. Furthermore, competitive events are directly related to classroom and laboratory instruction, thus providing a natural extension of learning through educative experiences in a social environment different from the traditional classroom.

A good example might be as follows:

A group of students who are enrolled in a food science course compete in the food science and technology CDE. Students learn technical content from their agriscience teacher in their course, meet once a week with a community member who is familiar with the event and works as a food safety specialist at the local chicken processing plant, and then travel to FFA competitions around the state where they compete against and interact with students from other FFA chapters.

Awards also incentivize learners to set and achieve developmental goals. In addition, they provide structure for teachers to organize learners' educational experiences. Like degrees of membership, awards are often earned by accomplishing a complementary set of tasks that draw on technical knowledge and skills as well as on leadership and interpersonal knowledge and skills. This is true for individual, team, and chapter awards. Some examples of awards include National Chapter Awards (awarded at the state and national levels), Agricultural Proficiency Awards (awarded at the local, state, and national levels), and Star Awards (awarded at the local, state, and national levels). Competitive events and awards can be unique and nontraditional educative experiences, but the teacher must plan diligently to achieve the desired student learning outcomes.

Conduct of chapter meetings is a common CDE offered by state associations and the national organization. The contest requires a team of students to conduct a meeting completing a specific list of tasks, according to prescribed parliamentary procedure. Conducting a meeting using parliamentary procedure is a valuable skill that requires a combination of technical knowledge and interpersonal skills to successfully accomplish the group's business. A teacher can utilize the conduct of chapter meetings contest to teach students correct parliamentary procedure through application and practice (experience). Learners apply their knowledge not only in the contest but also as FFA members directing the business of their local chapter and ultimately in adulthood as they engage in professional or civic organizations.

Events and Conferences

Events and conferences offer unique educational experiences. The National FFA Organization hosts numerous events and conferences. Some conferences are hosted by state associations with support from the National Organization. The National FFA Organization website and your state associations keep a current list of events and conferences offered. Examples include the National FFA Convention, Washington Leadership Conference (WLC), state or regional greenhand leadership camps, and many others.

Conferences offer another social environment in which teachers can organize student learning. FFA conferences are frequently focused on leadership and interpersonal skill development, though some may also focus on specific technical content. Learners return from conference experiences prepared with knowledge and skills to operate their local chapter. In this way, learners' knowledge and skills continue to build through personal and relevant experiences. For example, if a local chapter officer engaged in team-building workshops at a conference, they can apply that knowledge and skill to facilitate team-building activities that build cohesion and camaraderie among their chapter members. Alternatively, local FFA members may attend a conference where they learn about civil discourse and building consensus in group decision-making. They apply that knowledge to productively contribute to the development of their chapter's program of activities (POA). FFA conferences may last a few hours or several days. Each conference has specific learning outcomes for participants and incorporates a variety of activities and experiences to help learners achieve those outcomes. With careful planning, teachers can organize educational experiences that extend all learners' growth and development beyond the conference or event.

Diversity and Inclusion Efforts and Opportunities in FFA

FFA is intracurricular, which means it should be as accessible as its complementary components of the three-circle model (Classroom and Laboratory, Supervised Agricultural Experience, FFA). The teacher is responsible for ensuring efforts are made toward diversity and inclusion. FFA programs and activities can include all learners. The National FFA Organization continues to pursue diversity, inclusion, and equity throughout the organization as demonstrated through the National FFA Value Statements ratified by the delegates at the 2021 National FFA Convention.

Ratified FFA Value Statements

- We respect and embrace every individual's culture and experiences.
- We welcome every individual's contribution to advance our communities and the industry of agriculture.
- We cultivate an environment that allows every individual to recognize and explore their differences.
- We create leadership opportunities for every individual to enhance their personal and professional endeavors.

Additional efforts and ideas to increase accessibility are outlined in this section.

Affiliate program fees, paid by the chapter instead of by individuals, can reduce socioeconomic barriers to participation in FFA learning activities. Students' freedom to select from official dress options can reduce potential gender and cultural barriers. The National FFA Organization provides CDE and LDE resources in Spanish and offers a Spanish version of the Creed Speaking contest. Local chapters and agricultural classrooms should welcome and provide opportunities for all learners and strive to represent the demographics of the local school system and community.

FFA's history of supporting diversity and inclusion has evolved over the years from membership to leaders to policies and practices. The first female members were admitted in 1969. The first African American national leader was elected in 1994, and the first Latino national leader elected in 2002. In 2017, the first African American female national president was elected. The 2019 National Convention featured the first opening ceremonies ever conducted in Spanish. The National FFA Organization provides educator resources for teachers who wish to incorporate lessons on diversity and inclusion as well. You can find those resources on the National FFA Organization's website, www.ffa.org.

Teaching with FFA Educational Experiences

Teachers must intentionally plan and organize educational experiences. In SBAE, this includes FFA experiences. As illustrated by the three-circle model, classroom and laboratory instruction, SAE, and FFA are integrated and overlapping. Educational experiences can and should be organized to incorporate all three. Some examples of how teachers can integrate FFA into their teaching are presented below.

Mr. Brown is a SBAE teacher. He begins an instructional unit by requiring students to attend their local school board meeting to observe parliamentary procedure. Students then learn about proper meeting conduct through a series of lessons in class. Students compete in the conduct of chapter meeting CDE, gaining additional practice. Finally, students apply appropriate parliamentary procedure as they engage in the work of their local FFA chapter. Mr. Brown's approach illustrates one way teachers can use FFA to extend and deepen students' learning.

Josh has been enrolled in agricultural education courses since he was in middle school, where he earned the Discovery degree. One requirement of the Discovery degree was researching agricultural careers. Josh took a particular interest in landscape and horticulture careers. In high school, Josh enrolled in the plant science course, where he continued to learn more about horticulture. Josh earned his Greenhand degree and developed a plan for an SAE that continued to expand his interest in plants, horticulture, and landscape. He earned a spot on his chapter's plant identification CDE team and has started practicing for the floral CDE. Josh has a placement SAE delivering floral arrangements for the local flower shop. He enjoys interacting with customers in the shop and teaching them about the different plants and flowers. Josh's teacher has also asked him to comanage the school greenhouse and he serves on the annual plant sale committee for his local FFA chapter. Josh's SBAE experience exemplifies fully integrated classroom and laboratory instruction, SAE, and FFA.

The adviser of the Newberry FFA Chapter utilizes the POA to plan, organize and facilitate student learning experiences through FFA. Each year, the chapter officers lead members in developing a POA. The POA defines chapter goals and outlines plans to meet those goals. It is organized into three divisions: Growing Leaders, Building Communities, and Strengthening Agriculture. Each division contains five quality standards which allow for student committees to plan, prepare, and deliver activities in each area. Newberry FFA's POA is informed by community and local industry needs, considers the planned classroom and laboratory instruction, and addresses students' interests and needs. Students learn and practice a multitude of skills while developing the POA and while completing the planned activities throughout the year. Newberry FFA's approach to developing and carrying out a program of activities demonstrates how FFA, as a crucial component of the three-circle model, provides experiential education in SBAE programs. (For more information on Program of Activities, please see the resource guide on the National FFA website, www.ffa.org.)

Summary

1. Explain FFA as an intracurricular component of the complete agricultural education model.
2. Describe the structure and components of FFA.
3. Identify competencies taught through FFA activities.
4. Illustrate efforts and opportunities for diversity and inclusion in FFA.
5. Utilize FFA programs and activities as experiential learning opportunities for learners.

This chapter has described how FFA is an intracurricular component of a complete school-based agricultural education program thanks to its federal charter. FFA provides a three-tier structure and a multitude of activities and opportunities that support agriculture educators as they organize educative experiences for learners. These experiences allow learners to build upon their current knowledge and experience in a variety of social contexts, foster valuable adult-youth relationships, and challenge learners to integrate technical knowledge and skills with leadership and interpersonal knowledge and skills. Teachers must be intentional in selecting and organizing experiences that compliment and extend learning from the other two components of the three-circle model, classroom and laboratory instruction and SAE. In addition, teachers must continue to champion diversity and inclusion efforts, making SBAE accessible to all learners. Ultimately, when FFA is fully integrated with classroom/laboratory instruction and SAE, students received a robust and complete agricultural education that simultaneously supports the food, agriculture, and natural resources industry.

Learning Confirmation

1. Draw the three-circle model of agricultural education. Identify learning activities that exist in the center of the diagram where all three circles overlap as well as activities that exist where two of the three components overlap.
2. Outline an example program of activities. Explain how the FFA activities included support, extend, and enhance learning in the classroom and laboratory.
3. Identify a Career Development Event of your choice. List personal development and leadership competencies learners will practice simultaneously with the required technical content knowledge.

Applying the Content

1. Identify three Career Development Events that can be used in a local context to prepare FFA members for future careers in their community.
2. Identify three Leadership Development Events that can be used in a local context to teach leadership skills that FFA members can use in their community.
3. Prepare an introductory lesson to teach about FFA opportunities to new agricultural education students.

Reflective Questions

1. What challenges or barriers might teachers have to overcome to fully integrate FFA in their teaching?
2. How can agriculture teachers involve stakeholders, especially those in the community, to strengthen the connection between the classroom and laboratory, SAE, and FFA learning?
3. FFA offers a broad range of activities for teachers to incorporate. How can teachers identify the most effective and appropriate learning experiences for learners?

Glossary of Terms

- **experiential learning:** The process of learning by doing that involve hands-on experiences and reflection by learners
- **Supervised Agricultural Experience (SAE):** a planned and supervised program of experience-based learning activities that extend school-based instruction and enhance knowledge, skills, and awareness in agriculture and natural resources
- **school-based agricultural education (SBAE):** Systematic program of instruction in agriculture, food, and natural resources for K-12 students within a school program.
- **National FFA Organization (FFA):** A youth leadership organization that strives to make a positive difference in the lives of young people by developing their potential for premier leadership, personal growth, and career success through agriculture education
- **proficiency awards:** a program to recognize FFA members for their performance in supervised agricultural experience (SAE) programs; students can be recognized at the local, state, and national level
- **Kolb's Experiential Learning Theory:** Theory that concrete experience, reflective observation, abstract conceptualization and active experimentation form a four-stage process (or cycle) transformed into effective learning
- **Career Development Events (CDE):** Events sponsored through FFA where students compete in career areas to demonstrate their skill and expertise
- **Leadership Development Events (LDE):** Events sponsored through FFA where students compete in leadership areas such as public speaking and parliamentary procedure to demonstrate their skill and expertise
- **Program of Activities (POA):** A plan for executing events and activities typically developed by student leaders with guidance by the agriculture teacher
- **three-circle model:** A conceptual model for the total agricultural education program including classroom/laboratory instruction, FFA, and supervised agriculture experience

Resources

The **National FFA Organization website**, www.ffa.org, houses a plethora of resources for teachers, including resources for the classroom/laboratory, information on competitive events, awards, conferences, and other learning experiences, and advice for agricultural educators from agricultural educators.

FFA New Horizons is a magazine published quarterly by the National FFA Organization. An archive of issues can be found at https://archives.iupui.edu/handle/2450/2473 courtesy of Indiana University Purdue University Indianapolis University Library Ruth Lilly Special Collections & Archives.

The Agricultural Education Magazine, https://www.naae.org/profdevelopment/magazine/, is a professional journal published by the National Association of Agricultural Educators and provides an outlet for news, research, and other content appropriate for and pertinent to agricultural educators.

The ***Journal of Agricultural Education***, published by the American Association for Agricultural Education can be accessed at https://jae-online.org/index.php/jae. A rigorous peer-reviewed publication outlet for agricultural education faculty at higher education institutions, it is a source of science-based recommendations for agricultural educators. Some relevant articles include:

- Martin, M. J., & Kitchel, T. (2015). Advising an urban FFA chapter: A narrative of two urban FFA advisors. *Journal of Agricultural Education, 56*(3), 162–177. http://doi.org/10.5032/jae.2015.03162
- Rose, C., Stephens, C. A., Stripling, C., Cross, T., Sanok, D. E., & Brawner, S. (2016). The benefits of FFA membership as part of agricultural education. *Journal of Agricultural Education, 57*(2), 33–45. http://doi.org/10.5032/jae.2016.02033
- Shellhouse, J. A., Suarez, C. E., Benge, M., & Bunch, J. C. (2021). Mentoring mentality: Understanding the mentorship experiences of FFA officers. *Journal of Agricultural Education, 62*(1), 29–46. http://doi.org/10.5032/jae.2021.01029
- Wakefield, D., & Talbert, B. A. (2003). A historical narrative on the impact of the New Farmers of America (NFA) on selected past members. *Journal of Agricultural Education, 44*(1), 95–104. https://doi.org/10.5032/jae.2003.01095

The Friday Footnote, https://footnote.wordpress.ncsu.edu/, is a weekly blog written by Dr. Gary Moore, Professor Emeritus at North Carolina State University. His writing focuses on the history of agricultural education and rural America, including FFA, and includes a plethora of references to additional resources.

About FFA: https://www.ffa.org/about/

Agricultural Education Inside the Agrobox: https://www.ffa.org/ffa-new-horizons/agricultural-education-inside-the-agrobox/

Owl Chat: Educational Resources from National FFA: https://www.ffa.org/ffa/owl-chat-educational-resources-from-national-ffa/

SAE for All: Evolving the Essentials: https://saeforall.org/immersion-sae/

Town Hall Discusses Agricultural Education for All: https://www.ffa.org/agricultural-education-for-all/town-hall-to-discuss-agricultural-education-for-all/

Figure Descriptions

Figure 10.1: Three circle Venn diagram. Circle 1 represents classroom/laboratory. This is contextual, inquiry-based instruction and learning through an interactive classroom and laboratory. Circle 2 represents FFA. This is premier leadership, personal growth, and career success through engagement in FFA, PAS, or NYFEA programs and activities. Circle 3 represents SAE. This is experiential, service, and/or work-based learning through the implementation of a supervised agricultural experience program. All three circles meet in the center to represent school-based agricultural education.

Figure 10.2: Circular diagram. Concrete experience points to observations and reflections points to formation of abstract concepts and generalizations points to testing implications of concepts in new situations. Points back to concrete experience.

Figure References

Figure 10.1: Three Circle Model. Kindred Grey. 2023. Adapted under fair use from FFA, "The Three-Component Model." https://www.ffa.org/agricultural-education/

Figure 10.2: Kolb's experiential learning model diagram. Kindred Grey. 2023. Adapted under fair use from David A. Kolb, "Experiential Learning: Experience As The Source Of Learning And Development," 1984. https://www.researchgate.net/publication/235701029_Experiential_Learning_Experience_As_The_Source_Of_Learning_And_Development

References

Baker, M. A., Robinson, S., & Kolb, D. A. (2012). Aligning Kolb's Experiential Learning Theory with a comprehensive agricultural education model. *Journal of Agricultural Education*, 53(4), 1–16. https://doi.org/10.5032/jae.2012.04001

Committee on Agricultural Education in Secondary Schools. (1988). *Understanding agriculture: New directions for education.* The National Academies Press. https://doi.org/10.17226/766

Dewey, J. (1938). *Experience and education*. Simon & Schuster.

Kolb, D. A. (1984). *Experiential learning: Experience as the source of learning and development*. Upper Saddle River, NJ: Prentice Hall.

Roberts, T. G. (2003). An interpretation of Dewey's experiential learning theory. https://files.eric.ed.gov/fulltext/ED481922.pdf

The National FFA Organization. (2021). *About FFA*. https://www.ffa.org/about/

The National FFA Organization. (2019a, March 13). FFA Federal Charter. https://ffa.app.box.com/s/kpbbzar0hjej0i8odx9gg065vfow8boa

The National FFA Organization. (2019b, November 11). National FFA Constitution. https://ffa.app.box.com/s/m5882jqjl7cxhn5kr971xl57nib4r5le

Future Farmers of America Federal Charter, 36 U.S.C. §709. (2019). https://uscode.house.gov/view.xhtml?path=/prelim@title36/subtitle2/partB/chapter709&edition=prelim

11. SUPERVISED AGRICULTURAL EXPERIENCES

Eric Rubenstein

Setting the Stage

As we look to examine the final ring of the three-circle model in school-based agricultural education (SBAE), take a few moments to read this teacher's story about engaging students in Supervised Agricultural Experience (SAE) programs:

> *Every year it seems to get harder and harder. My students seem to be further removed from the ag industry and have little knowledge about how their food is produced. When I start my lessons on SAE at the beginning of each course, I continually worry about how I am going to find a meaningful program for each of my students. This past summer I had an opportunity to sit down and have an open and honest conversation with several other ag teachers and a teacher educator about my struggles with SAE in my classroom. Over the next hour, I learned that I was not alone but that each of us had different things that worked to engage some students. As I prepare to introduce SAE to a new group of students in a few weeks, I am excited for the first time to try something new.*

Take some time to reflect on the story shared by this twenty-two-year veteran agriculture teacher and how she has continued to push herself to learn and try new ideas regardless of the struggles she has had in the past. While SAE can seem overwhelming, being an agriculture teacher is about providing experiences for students to grow and develop. If we never lose sight of this goal, there are no challenges that we cannot overcome in the classroom. Take a few moments before you continue to write down a few ways that you plan to overcome challenges and work to keep a positive outlook on your tenure in the agricultural education profession.

Objectives

The purpose of this chapter is to introduce our readers to Supervised Agricultural Experience (SAE) programs. More specifically, the objectives are to:

- Discuss the purpose of SAE in SBAE.
- Discuss the struggles and concerns that exist within SAE development and implementation.
- Develop SAE programs for all students in SBAE.
- Discuss a model for SAE program development and implementation.
- Implement all components of SAE.

Introduction

As we begin to explore one of the experiential learning opportunities that students engage in while enrolled in a school-based agricultural education program, we must begin to investigate why students might benefit from these programs and how we as teachers can facilitate a learning environment where students feel safe to talk about the failures they experience. As teachers and members of society, we know that some of our most impactful learning experiences came from failures or struggles that we faced. One of the best ways we can authentically engage students in their own learning is through a project that is tailored to their personal interests and desires. Supervised Agricultural Experience programs provide students with authentic learning experiences that connect with personal interest areas with the vast agriculture, food, and natural resources system. Through this chapter, we will explore the history, benefits, struggles, and learning applications of Supervised Agricultural Experience programs within school-based agricultural education.

Overview of the Supervised Agricultural Experience Programs

What is SAE?

Agricultural educators across the United States utilize the principles of Supervised Agricultural Experience (SAE) programs to enhance the education quality and rigor of their programs. SAE is one of the core components of the three-circle agricultural education model that is used as the backbone of school-based agricultural education (SBAE). SAE is defined as "a planned and supervised program of experience-based learning activities that extend school-based instruction and enhance knowledge, skills, and awareness in agriculture and natural resources" (Barrick & Estepp, 2011). However, SAE participation has been continually decreasing over the past 50 years. To better understand the causes for this decrease, we must begin with how this teaching method came to be and the changes that have occurred over the past 110 years.

In 1908, Rufus Stimson was hired to serve as the director of Smith's Agricultural School in Massachusetts, where he was responsible for the educational program and operation of the school's farm (Stimson, 1915). During his time engaging with students, Stimson realized that there was a disconnect between the content being taught at the school and on the school farm and the student being able to transfer this knowledge to real-world issues taking place on their home farms. This phenomenon led Mr. Stimson to develop what he titled the project method, in which students engage in supervised programs of educational quality on their home farms to apply the content being taught in their classroom in real-world conditions. Over the years, the name of the program has been changed and renamed to meet the needs of the agricultural education profession, with the most recent name being Supervised Agricultural Experience. Due to the innovative nature of this teaching method, Stimson's approach was incorporated in the passage of the Smith Hughes Act of 1917, requiring all students enrolled in agricultural education to conduct an SAE program as part of their instruction. Students were required to have a "directed or supervised practice of agriculture either on a farm provided for by the school or other farm, for at least six months per year" (US House of Representatives, 1917). The federal legislation required teachers to provide supervision of these projects and submit annual reports, while on a twelve-month employment contact (US House of Representatives, 1917).

Since then, several other pieces of federal legislation have been passed and adopted, starting with the Vocational Education Act of 1963. This federal legislation removed the mandate for students to engage in an SAE and changed the wording that SAE may be provided to all students as a part of their SBAE instruction. Finally, in 1968 an amendment was passed to the Vocational Education Act that stripped all language related to SAE from federal legislation. Since this time, the removal of the federal requirement has decreased the utilization of SAE within SBAE programs across the United States. Until recently, many states did not have legislation related to agricultural education.

However, some states have begun passing legislation requiring students to be engaged in all three circles of the agricultural education three-circle model. For example, in April 2018, Georgia governor Nathan Deal signed into law the Green Agricultural Education Act, also known as Senate Bill 330. This act required that all students enrolled in an agricultural education course must be a member of the National FFA Organization and have an active SAE program as integral components of the course.

In addition to federal and state legislation, The National Council for Agricultural Education began an initiative in 2010 to rejuvenate SAE. The rejuvenation project was originally led by Dr. Kirby Barrick and examined the current status of SAE and suggestions for improvement to help SAE move into the twenty-first century. At the end of the initial work, lead to the development of sixteen tenants to SAE instruction and implementation (Barrick & Estepp, 2011). In 2018, this work led to the development of a new instructional strategy for SAE, SAE for All. Since 2018, this instructional technique has been slowly implemented across the United States.

SAE Development and Implementation

To help facilitate the growth of SAE in agricultural education, research has been conducted to examine the factors needed to develop and implement high-quality SAE programs for students regardless of their community demographics. Therefore, the following model was developed to help illustrate the necessary factors needed to ensure that SAE program development and implementation is successful. Five major factors emerged as essential components of the development and implementation process: (1) student-centered SAE programs, (2) committed teachers, (3) supporting surrounding "community," (4) shared experiences, and (5) joint supervision (figure 11.1).

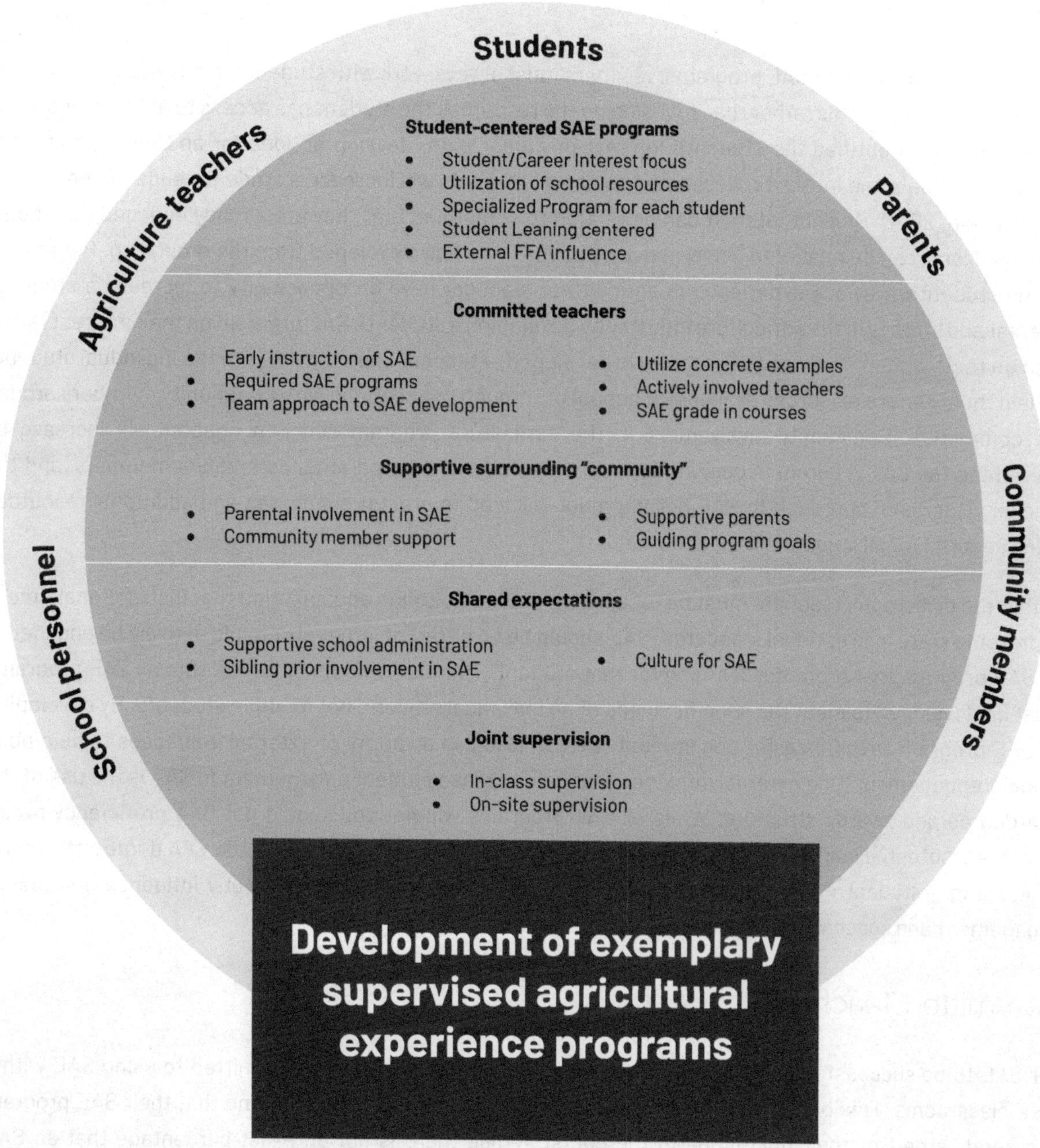

Figure 11.1. Model for the Development and Implementation of Exemplary Supervised Agricultural Experience Programs.

Student-Centered SAE Programs

When developing student SAE programs, teachers must always work with students to develop their areas of interest and career paths rather than focusing on the resources the student has access to at home or school. Once you have identified the student's interest or career path, develop an idea for an SAE program that meets their goals and interests. Research has shown that this will increase a student's engagement in SAE (Rubenstein, 2014). Student interest can be related to something that they are already engaged in at home or school, or it can be related to a new idea that needs to be fully developed from the ground up. Regardless of the student's previous experience in agriculture, teachers have an opportunity to increase a student's interest and literacy in the agriculture industry by developing a student's SAE program on their interest rather than on their resources. Once the program is developed, a teacher should work with the individual students to help them secure resources available at school or through connections with community members around the community. Community involvement in the agriculture program and FFA chapter will increase by developing the SAE program in conjunction with the student, their parents, community members, and the teacher. This team approach to SAE development will lead to greater successes and additional resources being secured for all students in the program.

Within the classroom, teachers must be focused on student learning and ensuring that instructional time is addressing state instructional standards. SAE should be no different, students should actively be engaged in an SAE program that promotes learning for that student. Therefore, we must realize that all SAE programs must be developed to meet the specific needs of an individual student, not multiple students. By developing an SAE program specifically for one student, we can also use a variety of external influences to help build student engagement. One external influence that can increase student engagement in SAE is the use of the FFA degree and awards structure. While all SAE programs will not and should not fit a proficiency award area, their potential learning and engagement in SAE can help them grow within the FFA degree structure. By keeping a student's SAE program focused on the student, teachers can greatly influence a student's engagement and success within their chosen SAE program.

Committed Teachers

For SAE to be successful in any program, the agriculture teacher must be committed to using SAE within their classrooms. This begins by requiring all students to be engaged and ensuring that their SAE program is graded as part of their final grade in the course. While there is not an exact percentage that an SAE program should account for in the overall course grade, it has been found to be important that there are graded components and that participation is tracked by the teacher. Furthermore, a teacher's instruction in SAE must begin within the first two weeks of the course to ensure that students begin the development and implementation process as early in the course as possible.

During this instruction, teachers should ensure that students have adequate time to complete the graded portions and needed hours of engagement before the end of the course. During the development phase, teachers should encourage students to talk to their parents, community members, local employers, and extended family members to help in the generation of an SAE program topic. When developing an SAE topic and program, teachers should utilize a team approach to ensure that the student has a large support network. Many teachers have shared that having periodic check-ins throughout the course helps students remain accountable for their engagement in SAE. During each of these check-ins, teachers need to remain actively involved in the students SAE programs by giving feedback to their SAE assignments, journal entries, and student involvement in their respective SAE programs. By remaining actively engaged in the students SAE program, teachers can provide the needed support to community members, parents, employers, and the students who are involved in the SAE development and implementation team.

Supportive Surrounding "Community"

As discussed in the last section, a supportive team or "community" is essential to the development of an exemplary SAE program. This team consists of the agriculture teacher(s), student, parents, community members, employer, and any other individuals who have had a significant role in assisting the student with the SAE program. One of the most important individuals in this process is the parent. Many parents want to be actively involved in the student's education experience; however, this sometimes is very difficult due to a lack of information provided to parents from teachers. Therefore, agriculture teachers must find a way to provide valuable SAE information to parents, so they are aware of the requirements of a student's SAE program. To facilitate active parental involvement, teachers should host an SAE night or send home a packet of information that describes the SAE program and gives ways for parents to support the program from home. This information packet or presentation should include the basics but not be overwhelming to the parent. Furthermore, community members should be actively engaged in the SAE program. Through personal connections with various community members, teachers can have informal conversations with many of these individuals to ensure that they are aware of the requirements and fully understand their role in the program. Be sure to set clear expectations with community members so they do not overtake the SAE program. As a team, the student and their support system should develop a set of program goals that they will utilize to move the program forward each year. These goals should be achievable by the student and appropriate for the given SAE program.

Shared Expectations

In some situations, students may have had a sibling or parent that was actively involved in agricultural education and may have a preestablished SAE program that will meet the class and program expectations. Students who have this type of opportunity can typically build on what was completed prior to their enrollment in the SBAE program and continually improve the SAE program over time. In the best-case scenario, siblings can work together in their SAE program to share ideas and make joint decisions that will benefit their SAE program and learning. Beyond the students, teachers must work to ensure that the school administration has a firm understanding of what SAE is, which helps to ensure their support of this fundamental portion of SBAE. To do this, teachers should demonstrate how students learn, make essential decisions, apply classroom knowledge, and continually build an exemplary SAE program that enhances their engagement in the agriculture curriculum. For example: students can give a presentation to school administration about the SAE programs, administrators can come to the agriculture classroom during SAE workdays to learn more about student programs, or even have administrators accompany the agriculture teacher on supervisory visits to students SAE programs. These activities have been found to increase administrators support of SAE programs within their schools.

Finally, a teacher must develop a "culture" for SAE within the SBAE program. A culture for SAE is not developed overnight and will be difficult to build from the ground up. Teachers must establish a firm expectation that all students will engage in an SAE and that there are no exceptions. Agriculture teachers who have established a "culture" for SAE have found that student engagement in SAE has increased significantly and that students know before they enroll in an SBAE course that they will be required to complete an SAE program as a portion of the agriculture course. Teachers have reported that this can take up to five years to establish (Rubenstein, 2014); however, once this has been successfully implemented, student engagement in SAE was established much quicker in the agriculture course. SAE culture is possibly the most important factor that influences the development and implementation of exemplary SAE programs and the completion of these programs by all students enrolled in the agriculture course.

Joint Supervision

There are two forms of supervision in SAE programs: on-site supervision and in-class supervision. Both forms of supervision are accepted and needed to ensure that all students engage in an SAE program. Over the past fifty years, an agriculture teachers' time has become stretched thinner and thinner so that they can accomplish more tasks than ever before, but in the same amount of time. Therefore, it is not expected that the same supervisory expectations for a teacher who taught fifty total students can be expected for a teacher who can now teach up to two hundred. Supervisory roles have changed, and need to continue to change and evolve, to meet the needs of both the student and the teacher. Therefore, it is recommended that all students receive in-class supervision through SAE workdays. These workdays should be completed at least every other week for approximately fifteen minutes to ensure that students are updating their records and that they have time to share concerns or issues about their SAE programs with their peers or teacher.

During this time, the teacher should try to talk with students about their SAE program and to help answer any questions or concerns they may have to ensure their continued success. For situations where a simple answer will not suffice or the teacher does not have enough information to answer the question, an on-site supervisory visit is needed. An on-site supervisory visit should be scheduled with the student and their parents and/or employer prior to the teacher's arrival. During the visit, the teacher should observe the student engaging in their SAE and then have adequate time to ask questions and provide the needed advice. This should also include time for the teacher to explain more about an SAE program with parents or employers to ensure they are adequately prepared to support the student in this vital learning environment.

Developing Student SAE Programs

When developing an SAE program, students must develop a program that fits their interests. While resources to conduct the SAE program are important, a student's interest in the program content is vital for the longevity of the program. As a part of the development process, parents and teachers must play an active role to help students create a program that is applicable and manageable. Students can have high or low aspirations when it comes to an SAE program; therefore, it is vital that the decision is not made solely by the student without any counsel from other adults. To assist students in the development process, Rubenstein and Thoron (2013) developed the SAE Dichotomous Key to help students think more about their personal interests in agricultural content areas, rather than their available resources. Rubenstein et al. (2022) redeveloped the instrument to meet the growing needs of the SAE for All program categories (see figures 11.2a-g).

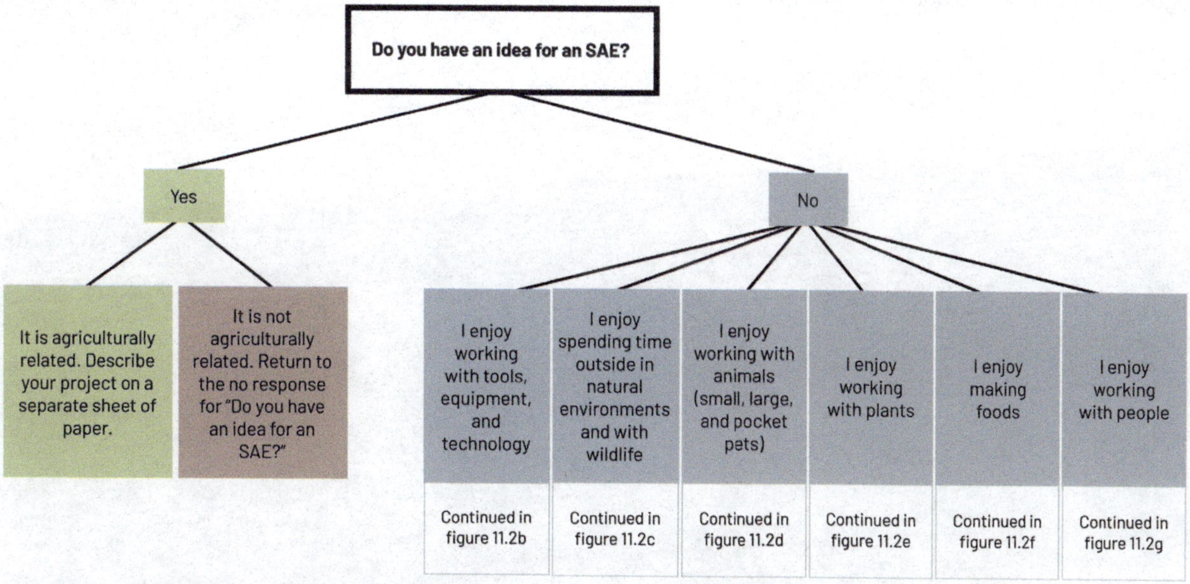

Figure 11.2a: The SAE Dichotomous Key (part 1).

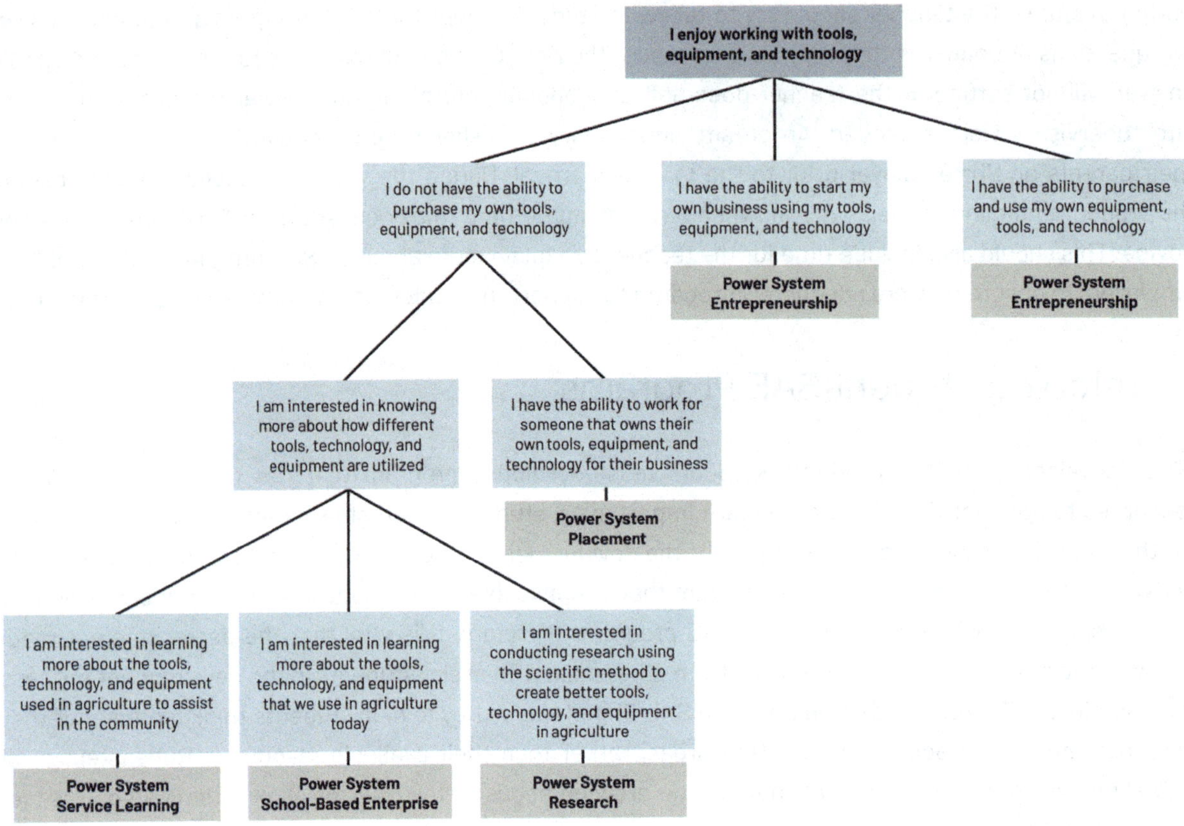

Figure 11.2b: The SAE Dichotomous Key (part 2).

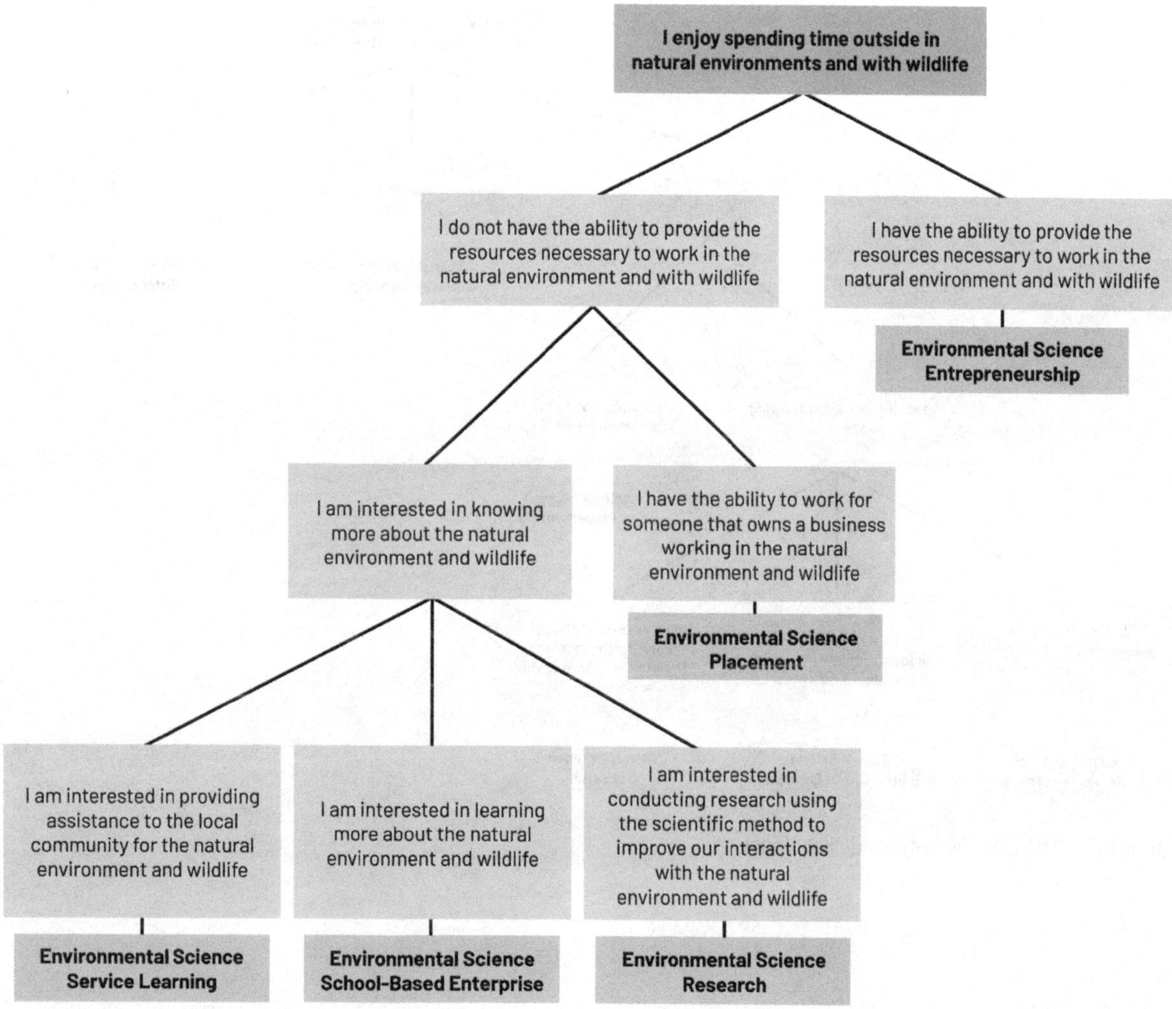

Figure 11.2c: The SAE Dichotomous Key (part 3).

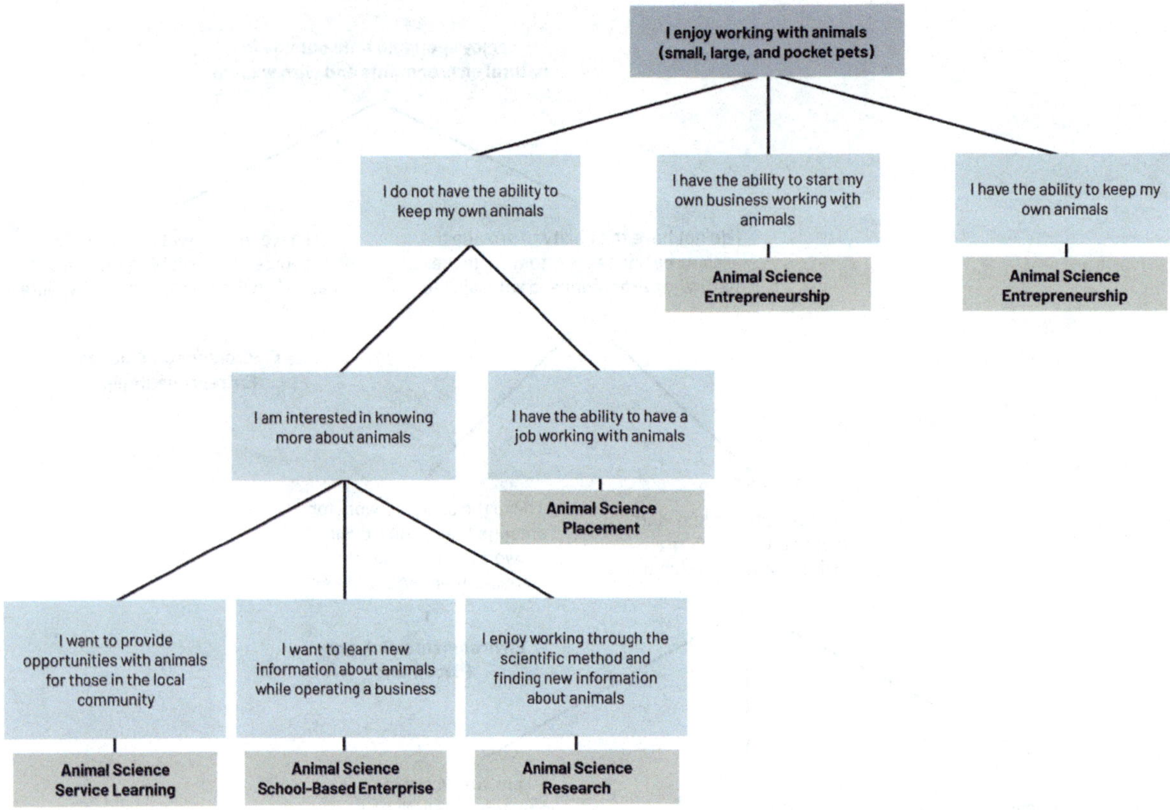

Figure 11.2d: The SAE Dichotomous Key (part 4).

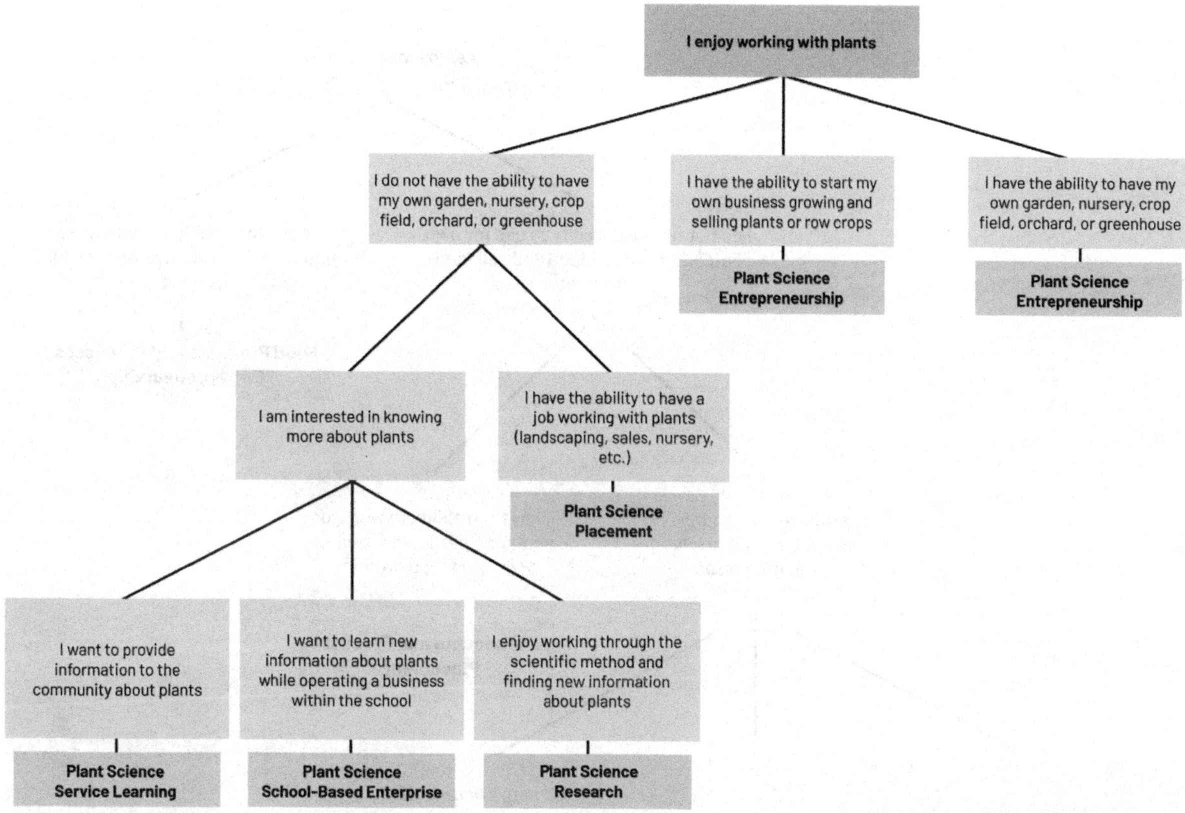

Figure 11.2e: The SAE Dichotomous Key (part 5).

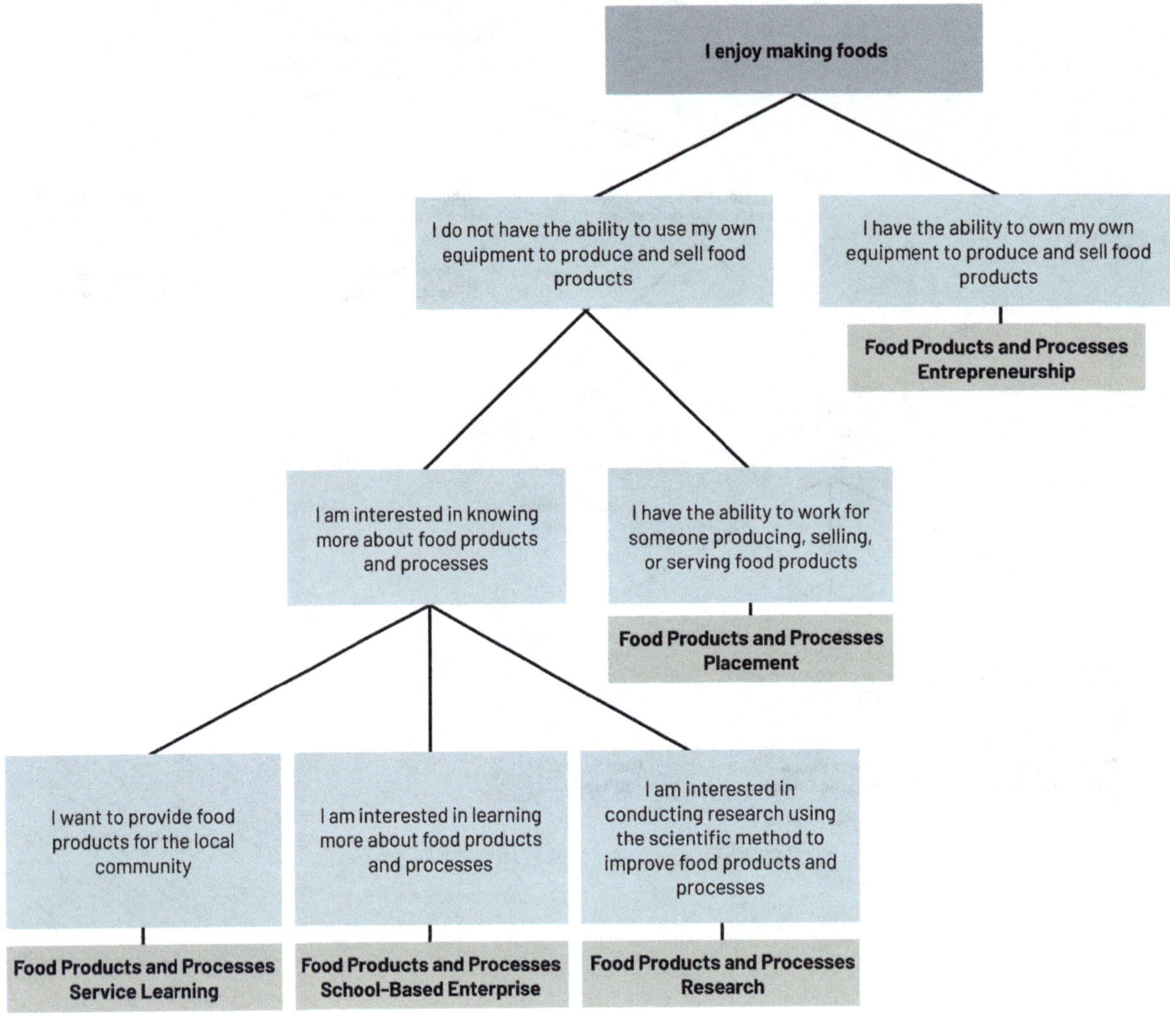

Figure 11.2f: The SAE Dichotomous Key (part 6).

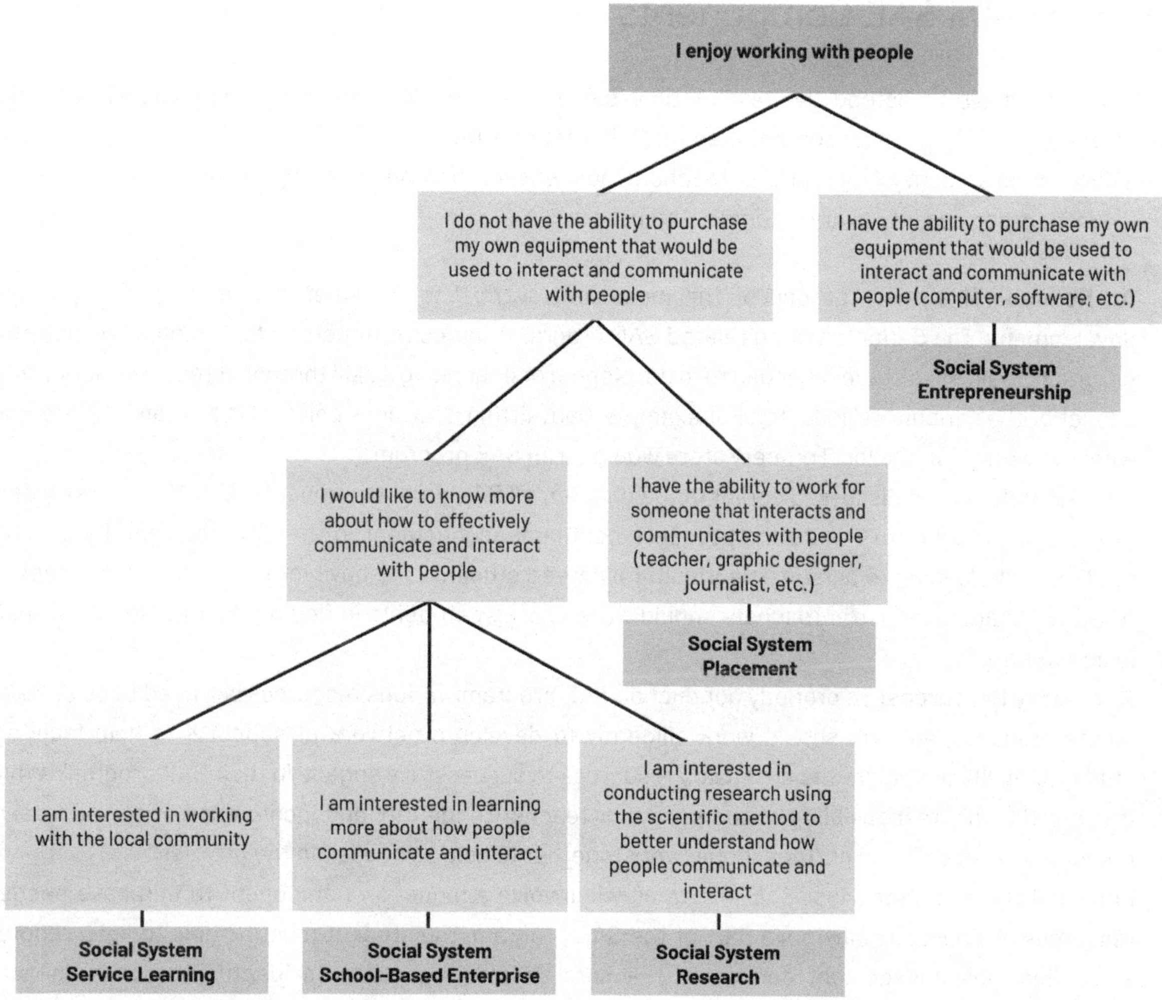

Figure 11.2g: The SAE Dichotomous Key (part 7).

Teachers should have students complete the dichotomous key independently. Unlike a traditional dichotomous key, there are multiple correct answers for a student on the SAE Dichotomous Key. Students should move through each possible question to determine whether they have a possible interest in each of the potential SAE programmatic categories. Once a student has selected all the programmatic categories of interest, students should begin to look at examples of different SAE programs that fit with in that SAE area. This activity can help many students develop a personalized SAE program that meets both their interests and available resources.

Fundamental SAE Components

SAE programs have withstood the test of time since it is the one area where agricultural education differentiates itself from other content areas with Career and Technical Education and core subject areas. When developing student SAE programs, teachers must ensure that the following fundamental components have been incorporated into every student's SAE program:

1. **Ag Related:** This might be one of the most controversial topics when it comes to SAE program development. The definition of ag related SAE programs varies from person to person. However, SAE programs must have students engaged in developing or enhancing a skill through direct interaction with a facet of the agricultural industry. For example, babysitting a farmer's children is NOT an SAE program whereas working in the local grocery store would be an SAE program.
2. **Interest Related:** As discussed earlier in the chapter, all SAE programs must be based on the student's interest in the topic to ensure the student's continual engagement in the SAE program. Even if the student isn't planning to pursue an agricultural career, students will have interests in different sectors of the agricultural industry. Teachers should work to assist students in finding their personal interests in agriculture.
3. **Necessary Resources:** To properly conduct an SAE program, various resources will need to be secured by the student. Teachers should work diligently to develop a network of supports to help facilitate students' ability to acquire the necessary resources to successfully engage in their SAE program. While it is not the sole responsibility of the parent or teacher to identify and locate each of the necessary resources, each program partner should work together to help as needed in the process.
4. **Financial Management:** Most SAE programs will involve a monetary component to purchase needed resources or financial gains. As a part of the SAE program, the student must be able to keep records of all financial management decisions. Teachers should help students identify the best financial management system for a student's program.
5. **SAE Decisions:** Most important of all, students must keep records on the decisions they make during their SAE programs. These decisions include purchasing new equipment for an enterprise, selling an animal, hiring a new employee, firing an employee, adding more homes to a lawn care business, and so forth. This must be recorded in a place that can be easily accessible to the student.

SAE Program Categories

Based on what a student is doing in their SAE program, they will need to select an SAE program category for their SAE program. Each program area will have different record keeping needs. For example: financial records will need to be kept for an entrepreneurship, placement, research, and school-based enterprise program, but financial records would not be necessary in a service-learning SAE program. Based on the student's program, teachers must help the student identify the correct category to ensure that they have all the necessary records for their SAE program, particularly if the student decides to apply for a degree or proficiency award within the National FFA Organization. Depending on the state in which you reside, different SAE program categories may be accepted by your State FFA staff. Below each of the SAE program categories in SAE for All will be described.

- **Entrepreneurship:** Students who conduct an entrepreneurship SAE own and operate their own business enterprise (this also includes show animals). Students will be responsible for acquiring the appropriate resources to operate their business and should provide the teacher with a business plan. The business plan should include the goods or services the student will provide to customers, a projected budget, and a marketing plan to solicit business. Teachers and parents will provide most of the supervision in this program.
- **Placement:** Students who conduct a placement SAE are engaged in some form of employment (paid or unpaid). During a placement SAE program, students will be given a list of job responsibilities, from the employer, that they must complete. These tasks should be agreed upon by the student and the employer, normally the teacher will only receive a list of the responsibilities from the student. Teachers and the employer should provide supervision, while the employer would provide all performance evaluations.
- **Research:** Students who conduct a research SAE program will utilize the scientific process to investigate phenomena within the agricultural industry. Students will typically work closely with their agriculture teacher, a college professor, or a specialist within their research area. Most times, students will need to secure lab space at the school to conduct their experiments and record their scientific data. During this SAE, the teacher and/or supervisor will provide much of the supervision.
- **School-Based Enterprise:** Students who conduct a school-based enterprise SAE program will engage in operating a business from the school. In many cases, school-based enterprises will require multiple students to work together on this SAE. Teachers should select one student to have a supervisory role, the supervisor should have experience working in this SAE and work closely with the agriculture teacher to make decisions regarding the enterprise. During this SAE the teacher will provide the supervision.
- **Service Learning:** Students who conduct a service-learning SAE program will work as an individual or with other students to conduct service projects the benefit local organizations or community. Teachers should be able to share names of possible organizations that may have service project needs within the local community. Teachers will provide much of the supervision for this SAE category.

SAE Supervision

The supervision of SAE programs historically fell solely on the agriculture teacher, which was mandated through the Smith Hughes Act of 1917. However, since the passage of the Vocational Education Act of 1963 and the subsequent amendment in 1968, supervision responsibilities have fallen on individuals other than the agricultural education teacher. These individuals include, but are not limited to, parents, community members, extension agents, employers, and even extended family members. Due to this shift in supervision responsibilities, it is important for the agriculture teacher to provide detailed instructions on the role of a supervisor in a student's SAE program. This information can be shared with supervisors through a written document or an in-person or online training session. Each agriculture teacher should determine the best approach for their individual communities.

Supervision of SAE can be conducted through two means: in person or in class. In-person supervision can be conducted by any program partner working with the student on their SAE program. During an in-person SAE visit, the supervisor should do the following things to help ensure the SAE is operating smoothly: watch the student engage in their SAE program, take pictures of the student engaged in their SAE, meet with the student and parent to answer questions, provide details to the parent about the SAE program requirements, and provide advice to the student regarding the progress of their SAE. When planning for an in-person supervisory visit, the teacher or program partner should always schedule a time with the student and their parents, a supervisory visit should never be a surprise. Most supervisory visits last forty-five minutes to an hour to ensure that the supervisor has adequate time to work with the student and their parents on their SAE program. Finally, any advice that is given to the student should always be based on science and previous experience. Teachers and program partners that do not have experience in the specific content area of the SAE should refrain from providing advice that will adversely affect the outcome of the SAE program, as this may lead to issues between the student or parent and the agricultural education program.

The second form of SAE supervision is in-class supervision. As noted earlier in the chapter, in-class supervision should be conducted during SAE work times that are provided to students on a biweekly basis. During these supervisory visits, students should be permitted to engage in conversation with other students who have similar SAE programs. During these conversations, the agriculture teacher should be readily available to answer questions and provide advice regarding specific issues that are occurring in a student's SAE program. From these supervisory visits, agriculture teachers can make informed decisions regarding which students require on-site supervisory visits. Use this time with students wisely to have them update records, prepare FFA applications, and engage in meaningful conversation regarding the progress or issues occurring within their chosen SAE programs.

Grading SAE Programs

All SAE programs should have a graded component to them, as they are an essential part of the three-circle model of school-based agricultural education. While there is not prescribed way as to how often an SAE program should be evaluated or what requirements should be examined, the agriculture teachers should create a plan and rubric that sets clear expectations for students within the SBAE program. This plan and rubric should be shared with students at the beginning of school and SAE programs should be evaluated more than one time through the course. The weight of this grade varies by teacher and should be agreed upon all teachers in an SBAE program. If teaching in a multiteacher program, all agriculture teachers in the SBAE program should meet each year to determine how SAE programs are going to be evaluated to ensure that all students have a similar experience and requirement regardless of who their teacher is that school year. These similar expectations and experiences will lead to a development of an SAE culture within the school and directly affect a student's willingness to participate in an SAE program in future years.

Connection to FFA

While an award in FFA should never be the primary way that an SAE is introduced to students, it can support the rationale as to why engagement in an SAE program is vital for all students in an SBAE program. Within the National FFA Organization, there are several ways that an SAE program can benefit a member. First, students must have a high-quality SAE program in order to earn the various degrees that are awarded through the organization. Each degree increases the SAE engagement requirements, encouraging students to engage in their SAE at a higher level each year that they are enrolled in the SBAE program. Furthermore, for students with exemplary SAE programs, the National FFA Proficiency Award Program allows for students to be recognized for their outstanding accomplishments within their SAE program. These awards are given in nearly fifty areas each year to recognize the specialized skills students have developed through involvement in their SAE program (National FFA Organization, 2023). As stated earlier, the National FFA awards programs or degrees should not be the sole reason for a student to engage in or design a specific SAE program; however, these opportunities can and should provide additional incentive for students to become actively engaged in their SAE programs.

Learning Confirmation

Throughout this chapter, we discussed the current status of SAE within SBAE programs across the nation. To address the concerns and struggles that teachers face with developing and implementing SAE programs, we explained practices that can be used to assist in engaging all students in these specialized programs and helping students develop a passion and interest in the Agriculture, Food, and Natural Resource Industry. As we wrap up this chapter, take some time to develop an implementation plan for how you will engage all students in your classrooms in an SAE program. As you develop your plan, take into consideration each of the five major factors needed to effectively develop and implement SAE programs and the community that you will be teaching in either during your student teaching experience or your first year teaching. To help you develop this implementation plan, take some time to practice using the SAE Dichotomous Key and to develop SAE plans or goals for the mock students listed below and complete the reflection questions at the end of the chapter. These two exercises will help you develop your knowledge about SAE, the SAE for All initiative, and the current struggles and concerns that are being faced by teachers in SBAE classrooms. When you complete the development of an SAE implementation plan be sure to keep a copy that you continue to adapt and improve as you grow and develop as a teacher and learn more about your community and changing students.

Application of Content

One of the most challenging parts of incorporating SAE programs for teachers is the development of individualized SAE programs for all students. To help in this process, use the following mock students and the SAE Dichotomous Key to practice developing an SAE program for each mock student.

Mock Student 1

Name: Christopher

Age: Fifteen (Freshman)

Potential Career Choice: Something with biology

Personality: Christopher is a very quiet young man. He is a good student that is always on time to class and has his work completed. He does not seem interested in getting a job; he would rather focus on playing sports for the high school team.

Home Location: Apartment Complex (less than three blocks from various businesses including a pet store and veterinarian)

Interests:

- Riding his bike and working out
- Hunting with his father
- Listening to music
- Walking his Golden Retriever
- Working in his patio garden at his apartment and school
- Attending his science classes

Parental Support: You have never met his parents. You believe that his parents are divorced.

Mock Student 2

Name: Julie

Age: Seventeen (Junior)

Potential Career Choice: No clue—thinks that she just wants to go work after school

Personality: Julie is very outgoing. She enjoys working with others and currently serves as the class president. She enjoys living from the earth and using natural organic products. Many times, she will discuss these ideas and her passion for this way of life in class, including a discussion of the new TV show "Alaska: The Last Frontier."

Home Location: Downtown (lives within walking distance of many businesses)

Interests:

- Skateboarding
- Hunting with her father
- Cooking for her family
- Making crafts and scrapbooking
- Going to yard sales and flea markets

Parental Support: Her parents live together. Her parents have not seemed very interested in the agricultural education program. Julie's parents do not seem to support or embrace her organic and living from the land lifestyle.

Mock Student 3

Name: Amy

Age: Sixteen (Sophomore)

Potential Career Choice: No clue

Personality: Amy is very quiet and does not share much with you. She seems to be very successful with written communication. However, she does not work well in a team.

Home Location: Rural subdivision, not a farm

Interests: Unsure—but seems to do well with her assignments, normally they are very creative

Parental Support: Amy's father is on your advisory council and her mother is a graduate from your program. Amy's mother is a nurse at the local hospital and her father works in telecommunication. Her parents are extremely involved in the agricultural education program; however, Amy does not share much about her agricultural interests with you. You have had several meetings with her parents, and they have told you that she enjoys drawing and using the computer.

Reflective Questions

After reading this chapter, please reflect on the following questions about SAE programs within SBAE.

1. How would you describe SAE's value to the three-circle model of agricultural education?
2. Utilizing the content of this chapter, what steps would you take to ensure that the five factors for developing and implementing SAE programs were employed in your program? Explain your steps.
3. Are a student's interests or their resources more important when developing an SAE program? Explain your answer.
4. What steps can be taken to incorporate parents, administration, school personnel, and community members into students' SAE programs?
5. What would be the outcome if a student did not have all the fundamental components for their SAE? Consider on both an individual student and the entire SBAE program.
6. How would you incorporate SAE into student grading?
7. What criteria would you use (or provide to other supervisors) when supervising a student's SAE program?
8. In your opinion, what is the role of the agriculture teacher in a students' SAE program?
9. How would you prioritize student interest and motivation as an incentive for students to develop an SAE program?
10. What would you suggest to a first-year agriculture teacher to effectively implement SAE into their SBAE program?

Glossary of Terms

- **Entrepreneurship SAE:** An SAE program where students own and operate their own business enterprise
- **experiential learning:** – A four-phase process of learning through an experience
- **National FFA Organization (FFA):** A youth leadership organization that strives to make a positive difference in the lives of young people by developing their potential for premier leadership, personal growth, and career success through agriculture education
- **joint supervision:** Supervision of a student SAE both in class and on site by an agriculture teacher or other community member
- **Placement SAE:** An SAE program where students are in engaged in some form of employment (paid or unpaid)
- **proficiency award:** Award to honor FFA members who, through SAE, have developed specialized skills that they can apply toward their future career
- **Research SAE:** An SAE program where students will utilize the scientific process to investigate phenomena within the agricultural industry
- **Rufus Stimson:** An American educator with great impact on agriculture education with his development of project method of learning
- **SAE for All:** An initiative developed by the National Council for Agricultural Education to help rejuvenate the utilization of SAE in SBAE
- **School-Based Enterprise SAE:** An SAE program where a student will engage in operating a business from the school
- **Service-Learning SAE:** An SAE program where students will work as an individual or with other students to conduct service project that benefit local organizations or community
- **Smith Hughes Act of 1917:** Provided federal aid to the states for the purpose of promoting precollegiate vocational education in agricultural and industrial trades
- **student interest:** The inclination of the student toward a particular subject in which (s)he is easily able to connect
- **Supervised Agricultural Experience (SAE):** a planned and supervised program of experience-based learning activities that extend school-based instruction and enhance knowledge, skills, and awareness in agriculture and natural resources
- **Vocational Education Act of 1963:** Federal legislation that provided grants to states to maintain, improve, and develop vocational-technical education programs

Figure Descriptions

Figure 11.1: School personnel, agriculture teachers, students, parents, and community members all contribute to the development of exemplary supervised agricultural experience programs. The following 5 things also contribute. (1) Student-centered SAE programs. This includes student/career interest focus, utilization of school resources, specialized program for each student, student learning centered, and external FFA influence. (2) Committed teachers. This includes early instruction of SAE, required SAE programs, team approach to SAE development, utilize concrete examples, actively involved teachers, and SAE grade in courses. (3) Supportive surrounding community. This includes parental involvement in SAE, community member support, supportive parents, and guiding program goals. (4) Shared expectations. This includes supportive school administration, sibling prior involvement in SAE, and culture for SAE. (5) Joint supervision. This includes in-class supervision and on-site supervision.

Figure 11.2a: Flow chart. Do you have an idea for an SAE? If yes, either (1) It is agriculturally related. Describe your project on a separate sheet of paper; or (2) It is not agriculturally related. Return to the original question and select no. If you select "no" for the question "Do you have an idea for an SAE?", there are 6 subsequent boxes with answers to choose from. They are (1) I enjoy working with tools, equipment, and technology, (2) I enjoy spending time outside in natural environments and with wildlife, (3) I enjoy working with animals (small, large, and pocket pets), (4) I enjoy working with plants, (5) I enjoy making foods, and (6) I enjoy working with people. Each of these options are expanded on in the following 6 figures.

Figure 11.2b: Flow chart. You enjoy working with tools, equipment, and technology. If you have the ability to purchase and use your own equipment, tools, and technology, then the outcome is Power System Entrepreneurship. If you have the ability to start your own business using your tools, equipment, and technology, then the outcome is Power System Entrepreneurship. If you do not have the ability to purchase your own tools, equipment, and technology AND you have the ability to work for someone that owns their own tools, equipment, and technology for their business, then the outcome is Power System Placement. If you do not have the ability to purchase your own tools, equipment, and technology AND you are interested in knowing more about how they are utilized, there are three options: (1) If you are interested in learning more about the tools, technology, and equipment used in agriculture to assist in the community, then the outcome is Power System Service Learning; (2) If you are interested in learning more about the tools, technology, and equipment that we use in agriculture today, then the outcome is Power System School-Based Enterprise; (3) If you are interested in conducting research using the scientific method to create better tools, technology, and equipment in agriculture, then the outcome is Power System Research.

Figure 11.2c: Flow chart. You enjoy spending time outside in natural environments and with wildlife. If you have the ability to provide the resources necessary to work in the natural environment and with wildlife, then the outcome is Environmental Science Entrepreneurship. If you do not have the ability to provide the resources necessary to work in the natural environment and with wildlife AND you have the ability to work for someone that owns a business working in the natural environment and wildlife, then the outcome is Environmental Science Placement. If you do not have the ability to provide the resources necessary to work in the natural environment and with wildlife AND you are interested in knowing more about it, there are three options: (1) If you are interested in providing assistance to the local community for the natural environment and wildlife, then the outcome is Environmental Science Service Learning; (2) If you are interested in learning more about the natural environment and wildlife, then the outcome is Environmental Science School-Based Enterprise; (3) If you are interesting in conducting research using the scientific method to improve our interactions with the natural environment and wildlife, then the outcome is Environmental Science Research.

Figure 11.2d: Flow chart. You enjoy working with animals (small, large, and pocket pets). If you have the ability to start your own business working with animals, the outcome is Animal Science Entrepreneurship. If you have the ability to keep your own animals, the outcome is Animal Science Entrepreneurship. If you do not have the ability to keep your own animals AND you have the ability to have a job working with animals, the outcome is Animal Science Placement. If you do not have the ability to keep your own animals AND you are interested in knowing more about animals, there are three options: (1) If you want to provide opportunities with animals for those in the local community, then the outcome is Animal Science Service Learning; (2) If you want to learn new information about animals while operating a business, then the outcome is Animal Science School-Based Enterprise; (3) If you enjoy working through the scientific method and finding mew information about animals, then the outcome is Animal Science Research.

Figure 11.2e: Flow chart. You enjoy working with plants. If you have the ability to start your own business growing and selling plants or row crops, then the outcome is Plant Science Entrepreneurship. If you have the ability to have your own garden, nursery, crop field, orchard, or greenhouse, then the outcome is Plant Science Entrepreneurship. If you do not have the ability to have your own garden, nursery, crop field, orchard, or greenhouse AND you have the ability to have a job working with plants in landscaping, sales, nursery, etc., then the outcome is plant Science Placement. If you do not have the ability to own your own garden, nursery, crop field, orchard, or greenhouse AND you are interested in knowing more about plants, then there are three options: (1) If you want to provide information to the community about plants, then the outcome is Plant Science Service Learning; (2) If you want to learn new information about plants while operating a business within the school, then the outcome is Plant Science School-Based Enterprise; (3) If you enjoy working through the scientific method and finding new information about plants, then the outcome is Plant Science Research.

Figure 11.2f: Flow chart. You enjoy making foods. If you have the ability to own your own equipment to produce and sell products, then the outcome is Food Products and Processes Entrepreneurship. If you do not have the ability to use your own equipment to produce and sell food products AND you have the ability to work for someone producing, selling, or serving food products, then the outcome is Food Products and Processes Placement. If you do not have the ability to use your own equipment to produce and sell food products AND you are interested in knowing more about food products and processes, then there are three options: (1) If you want to provide food products for the local community, then the outcome is Food Products and Processes Service Learning; (2) If you are interested in learning more about food products and processes, then the outcome is Food Products and Processes School-Based Enterprise; (3) If you are interested in conducting research using the scientific method to improve food products and processes, then the outcome is Food Products and Processes Research.

Figure 11.2g: Flow chart. You enjoy working with people. If you have the ability to purchase your own equipment that would be used to interact and communicate with people (computer, software, etc.), then the outcome is Social System Entrepreneurship. If you do not have the ability to purchase your own equipment that would be used to interact and communicate with people AND you have the ability to work for someone that interacts and communicates with people (teacher, graphic designer, journalist, etc.), then the outcome is Social System Placement. If you do not have the ability to purchase your own equipment that would be used to interact and communicate with people AND you would like to know more about how to effectively communicate and interact with people, then there are three options: (1) If you are interested in working with the local community, then the outcome is Social System Service Learning; (2) If you are interested in learning more about how people communicate and interact, then the outcome is Social System School-Based Enterprise; (3) If you are interested in conducting research using the scientific method to better understand how people communicate and interact, then the outcome is Social System Research.

Figure References

Figure 11.1: Model for the Development and Implementation of Exemplary Supervised Agricultural Experience Programs. Kindred Grey. 2023. CC BY 4.0. Adapted from Eric D. Rubenstein, "Exemplary supervised agricultural experience programs in rural secondary schools," 2014.

Figure 11.2a: The SAE Dichotomous Key (part 1). Kindred Grey. 2023. CC BY 4.0. Adapted from Eric D. Rubenstein, "Updating the essentials: SAE for all through a dichotomous key," 2022.

Figure 11.2b: The SAE Dichotomous Key (part 2). Kindred Grey. 2023. CC BY 4.0. Adapted from Eric D. Rubenstein, "Updating the essentials: SAE for all through a dichotomous key," 2022.

Figure 11.2c: The SAE Dichotomous Key (part 3). Kindred Grey. 2023. CC BY 4.0. Adapted from Eric D. Rubenstein, "Updating the essentials: SAE for all through a dichotomous key," 2022.

Figure 11.2d: The SAE Dichotomous Key (part 4). Kindred Grey. 2023. CC BY 4.0. Adapted from Eric D. Rubenstein, "Updating the essentials: SAE for all through a dichotomous key," 2022.

Figure 11.2e: The SAE Dichotomous Key (part 5). Kindred Grey. 2023. CC BY 4.0. Adapted from Eric D. Rubenstein, "Updating the essentials: SAE for all through a dichotomous key," 2022.

Figure 11.2f: The SAE Dichotomous Key (part 6). Kindred Grey. 2023. CC BY 4.0. Adapted from Eric D. Rubenstein, "Updating the essentials: SAE for all through a dichotomous key," 2022.

Figure 11.2g: The SAE Dichotomous Key (part 7). Kindred Grey. 2023. CC BY 4.0. Adapted from Eric D. Rubenstein, "Updating the essentials: SAE for all through a dichotomous key," 2022.

References

Barrick, R. K., & Estepp, C. M. (2011). Experience programs in agriscience education: From projects to SAEPs and beyond. *Agricultural Education Magazine, 83*(4), 26–27.

National FFA Organization. (2023). *Agricultural Proficiency Awards*. National FFA Organization. https://www.ffa.org/participate/awards/proficiencies/

US House of Representatives. (1917). *Annual report of the federal board for vocational education*. Government Printing Office.

Rubenstein, E. D. (2014). *Exemplary supervised agricultural experience programs in rural secondary schools* [Doctoral Dissertation]. University of Florida.

Rubenstein, E. D., Scott, J. D., & Thoron, A. C. (2022, May 16-19). *Updating the essentials: SAE for all through a dichotomous key.* Poster presented at the 2022 National AAAE Conference in Oklahoma City, OK.

Rubenstein, E. D., & Thoron, A. C. (2013, May). A dichotomous key: Helping students find their path in SAE. *Proceedings of the 2013 National Agricultural Education AAAE Research Conference, Columbus, OH,* 7-10. http://aaaeonline.org/uploads/allconferences/5-16-2013_464_2013AAAE_PosterProceedings.pdf

Stimson, R. W. (1915). The Massachusetts home project plan for vocational agricultural education. *The School Review, 23*(7), 474–478. http://www.jstor.org/stable/1076877

12. EFFECTIVE USE OF THE AGRICULTURE LABORATORY ENVIRONMENT TO SUPPORT STUDENT LEARNING

Hannah H. Scherer

Setting the Stage

Most agricultural education programs have both classroom and laboratory facilities. Laboratories for agricultural instruction are not there out of tradition. Rather, laboratories are a crucial site of powerful teaching and learning in agriculture. Many physical spaces can be considered a laboratory environment; the defining characteristic is that students are using authentic resources in hands-on approaches. Within this definition, a traditional classroom can also be considered a laboratory environment if the teacher brings in materials. Consider the following scenario to help you start thinking about this topic:

Picture yourself as a new teacher getting to know the spaces in your program. You can already see how you will arrange the desks in your classroom and utilize the technology to support student learning, but you are really excited to get the students into the laboratory space. Do you have a greenhouse? A land lab? A woodworking shop? A biotechnology lab? What else? Picture it in your mind. If you need some ideas, see Appendix 12A for a list of possible in-school laboratory facilities by career pathway.

You know that you want to give the students plenty of hands-on time in the lab, but you also know that powerful teaching requires intentional planning for student engagement, inclusivity, and assessment. You start to wonder ... How do I address these elements in the lab? How do I decide when to teach in the lab and when to teach in the classroom? How does what students do in the lab relate to their learning in the classroom? How do I ensure that learning is happening when students are in the lab? What other questions do you have?

Objectives

By the end of this chapter, you will be able to:

- Differentiate among principal ways the laboratory environment can be used to support student learning.
- Design units of instruction that effectively utilize the laboratory environment to meet student learning objectives.
- Identify unique considerations in planning, delivery of instruction, and assessment in the laboratory environment.

Introduction

Just like in the classroom, powerful teaching in the laboratory environment involves backward design. Teachers must carefully consider the learning goals they have for their students when they make decisions about how to organize the learning activities in and outside the laboratory environment. While not an exhaustive list, this chapter presents three principal ways that teachers can use the laboratory environment to support student learning. Each mode is grounded in a different theoretical perspective on learning and supports different types of outcomes for students. Each will be appropriate at different times and can be used in combination when planning a laboratory-based course.

Important Note about Safety, Inclusion, and Laboratory Management!

An essential element of teaching in any laboratory environment is providing for safety instruction and monitoring students for safe practices throughout any lesson. Every agricultural laboratory has many dangerous areas and situations. It is crucial that students learn to work in their environment safely and that safety is considered throughout the planning process. Every facility is different (see Appendix 12A) and comprehensive coverage of safety and laboratory management is outside of the scope of this textbook. You will need to learn about safety instruction along with other important topics such as supervising students in the laboratory, managing tools and equipment, and establishing clean-up routines through working with a mentor, seeking out industry resources, and/or taking an agricultural laboratory management course. Equally important is to make every effort to include all students in the laboratory environment when it is safe to do so. Working with special education professionals in your building to modify activities to accommodate student needs and increase physical accessibility of your laboratory spaces is important to ensure that you meet the standard of inclusive learning environments. Staying up to date with current safety standards will also be part of your ongoing professional learning throughout your career. Potential mentors in this area include advisory board members, industry professionals, other teachers, and retired teachers. Some introductory considerations for safety instruction and laboratory management are provided in Appendix 12B.

Along with the design considerations for supporting student learning, it is essential that you ask yourself the following questions and seek out answers BEFORE any students enter the laboratory environment.

- What safety tests are students required to pass before using the tools and equipment included in this unit?
- What are my expectations for student behavior in the laboratory environment? How will I ensure that my expectations are clearly communicated and consistently reinforced? Have I adequately accounted for cultural differences among my students in setting and communicating my expectations? What are the consequences for a student who does not meet these expectations?
- What accommodations are required to ensure that all students have an opportunity to participate in the laboratory environment? Can I provide these accommodations, or do I need to seek additional support or guidance from a resource teacher? Who should I be collaborating with to meet the needs of all of my students?
- What is my plan for monitoring student behavior in the laboratory environment? How will I handle a student that poses a safety risk to themselves or others?
- Are there sufficient resources for all students to be doing the same thing, or do I need to plan for stations? If there are stations, what will they be and how will I ensure that all students are on task?
- Have I ensured that all of the tools and equipment students will need are in good working order?

Background: Science and Theory

An overview of direct instruction, experiential learning, and discovery learning is presented here as a foundation for the remainder of the chapter.

Direct Instruction for Technical Skill Development

Within direct instruction , the laboratory is viewed as a place where students learn and practice the technical skills needed for agricultural occupations. This learning happens through direct instruction (Hunter, 1982; Dell'Olio & Donk, 2007) that scaffolds psychomotor skill development as students work toward the independence that they will need to use these skills in the future. This teacher-directed mode is grounded in information processing and behaviorism, which help the teacher carefully consider how they organize instruction to best support student development of well-defined skills safely and to industry standards. When applied to psychomotor learning outcomes in agricultural education, the laboratory environment enables the teacher to use authentic tools and materials in carefully planned demonstrations to present the skill, design hands-on opportunities for guided and independent practice, and reinforce the learning through distributive practice in subsequent activities (Osborne, 1986).

Experiential Learning for Generalized Understanding of Real-World Processes

Within experiential learning, the laboratory is viewed as the site of authentic experience with real-world processes in agriculture. These experiences form the basis for experiential learning, which happens through a process of guided reflection and meaning-making to build toward a more generalized understanding of what happened that can then be used in subsequent experiences. Over time, this repeated process of experience, reflection, meaning-making, and applying new understandings (the "learning spiral" [Kolb, 2015]) can lead to a deeper ability to flexibly apply learning in practice. This mode is grounded in Kolb's experiential learning theory, which helps the teacher consider any experience in the laboratory environment (successful or not) as having the potential to contribute to student learning. When applied to process-oriented learning outcomes in agricultural education, the laboratory environment allows the teacher to provide real-world experiences in agriculture that are not possible in the classroom.

Discovery Learning for Coconstruction of Solutions to Real-World Problems

Within discovery learning, both the classroom and the laboratory are viewed as integrated resources to support students as they address real-world questions and problems in agriculture. This happens through student-centered instructional models grounded in inductive reasoning, such as problem-based learning, design-based learning, and inquiry-based learning. These models are grounded in social constructivism, which compels the teacher to design learning experiences in which students work together to address issues that matter to them so that they can develop their own solutions and explanations (Schunk, 2012). Within the context of instructional models that support problem solving and the development of higher-order thinking skills, the laboratory environment provides authentic resources (tools, equipment, spaces, etc.) for students to use in their efforts. This allows for student discoveries, designs, and solutions to be more realistic and relatable to the agricultural industry and/or their own lives.

Designing for Powerful Teaching in the Laboratory Environment

Deciding what, how, and when to teach in the laboratory involves decisions at all scales, driven by the question of "why?" Everything presented in this book about effective teaching, evaluation, and inclusion apply in the laboratory as much as they do in the classroom, but there are unique characteristics about the laboratory environment that require additional consideration. In this section, each mode described above will be illustrated with examples and further guidance for designing instruction within that mode.

Meet the Carroll County High School Agriculture Department

Vignettes in this chapter will feature ways in which teachers at Carroll County High School (CCHS) design for effective teaching in the laboratory environment. CCHS is a public, comprehensive high school located in a rural southwest Virginia. The school comprises grades nine through twelve and has approximately 1,150 students. Farm production and agriculture are the largest economic sector in Carroll County. CCHS's agriculture department has a STEM Lab for Agriculture equipped with the latest biotechnology research equipment, a dedicated sixty-five-acre farm, and an agricultural mechanics facility. The agriculture department serves students with a diverse range of career and postsecondary educational goals.

Designing Direct Instruction for Technical Skill Development

Direct Instruction in Action at CCHS

In her floriculture class, Ms. Sarah Jo Jones often transforms her classroom into a laboratory environment to teach floral design principles. She designs hands-on direct instruction lessons that build toward students independently designing floral arrangements for local clients. Today, she is teaching about balance in floral design. Her objective is for students to be able to construct a vase arrangement that achieves symmetrical balance. She has set out vases, stems, and greenery on her table at the front of the room and each student has the same materials at their station. There are additional materials at the front of the room. Ms. Jones begins the lesson with asking students if they have learned about the concept of balance in other settings, such as art class. She then announces that this will be the focus of class today and states the objective. She explains the concept of symmetrical balance and why it matters in floral design. Ms. Jones builds an arrangement to demonstrate how she uses the materials she has to achieve balance in her arrangement, pausing at key stages to have students do the same thing. While students are working, she circulates and gives each student feedback. After everyone has an arrangement, she leads a discussion to make sure students understand why she made the choices she did and how they can achieve balance in their work. Ms. Jones then directs students to gather additional materials from the front of the room and continue to build out their arrangement with the goal of maintaining symmetrical balance. Students submit a photo of their final arrangement to Ms. Jones for assessment. At the end of class, Ms. Jones reminds students that they will be using this concept along with other floral design principles when they work with their wedding client later in the semester. Finally, she instructs them to follow standard clean-up procedures before dismissal.

This vignette illustrates how Ms. Jones uses direct instruction in order to support technical skill development through engaging the students' interest, modeling the technique, and giving students a chance to practice with close guidance and feedback; they will continue to develop this psychomotor skill through opportunities for independent practice. Through evaluation, she can determine the level of skill development for each student and continue to support them through distributed practice of the skill within the context of real-world application. Considerations for each stage of direct instruction in the laboratory environment are as follows:

Focus Activity

A brief focus activity should draw students in to the lesson for the day. Safety is an important aspect of transitioning into the session for the day. Ensuring that students are focused and prepared for entering the laboratory environment is a daily task and can be supported through establishing routines as described in Appendix 12B.

Stating Objectives and Providing Rationale

For psychomotor skill development, performance objectives are most appropriate. Situating the objectives for the lesson in the context of what students are already able to do and how they will use the new skill in the future is an important step in direct instruction. Teachers can use Universal Design for Learning to individualize performance objectives to meet unique needs of students in their classes.

Presenting Content and Modeling

Modeling through a demonstration is the primary mode of content delivery for psychomotor skill development in the laboratory environment. Typically, the teacher will break down the skill into steps, using narration to explain how and why they are doing what they are doing. Careful planning for how this demonstration is sequenced, what the teacher does and says for each step, and how students will be positioned to best see what the teacher is doing are essential.

Checking for Understanding

Checking for understanding should happen throughout the demonstration. Before moving on to the next step, ask the class to answer questions related to what was just demonstrated, including any safety measures that are important to highlight. For safety reasons, making sure that students are able to explain what they need to do is often an important precursor to hands-on practice.

Guided Practice

Guided practice can be integrated with the demonstration or happen after the demonstration is complete. In some cases, the demonstration will happen in the classroom and then the class will move to a laboratory space for guided practice. In other cases, the demonstration will take place in the same space where students will practice the skill. When it is safe to do so, the teacher can have students follow along with a demonstration using authentic materials. In this case the teacher pauses at each key step and circulates to check that students have completed the task accurately before moving on, completing several cycles of modeling, checking for understanding, and guided practice within a single demonstration. Throughout guided practice, the teacher circulates, monitoring the individual actions of the students, providing one-on-one feedback and corrections, and bringing the whole class back together as needed for reteaching to address common errors or challenges. By the end of guided practice, the teacher is confident that each student is ready to practice the skill on their own.

Independent Practice

In this stage, students are given the opportunity to practice the skill on their own. This is important for psychomotor skills that they will need to use with fluency in subsequent projects or laboratory activities. Ensuring that each student can perform the skill correctly and safely on their own is an important precursor to independent practice. Managing student behavior and access to resources during independent practice requires careful planning (Appendix 12B).

Closure

Through closure, the teacher reinforces the need for the skill that was taught and previews for the students how they will use it in the future. Following closure of the lesson, students are dismissed to complete clean-up procedures before leaving the laboratory environment for the day.

Distributed Practice

Teachers of agriculture do not teach psychomotor skills independent of real-world application. Once students have developed the technical skills they need in a particular area, such as woodworking, they combine these skills and continue to practice each one as they apply them in the context of authentic projects and activities in the laboratory environment. Such projects can be designed as summative assessments of unit-scale transfer goals alongside evaluation of performance objectives related to individual skills.

Check for Understanding

Review the vignette describing Ms. Jones's floral design class. Can you identify what she did for each stage of direct instruction described here?

Designing Experiential Learning for Generalized Understanding of Real-World Processes

Experiential Learning in Action at CCHS

As part of a grant to test out production methods for extending the growing season, students at Carroll County High School work with Dr. Randy Webb to grow raspberries in high tunnel greenhouses on their land lab. Yesterday, Dr. Webb's agricultural production class worked in the greenhouses, harvesting ripe berries for sale. Each day they are in the greenhouses, students make an entry in their logbook recording what they did and observations they made. Today in the classroom, Dr. Webb leads a lesson on the life cycle of raspberry plants. His objective is for them to be able to use their knowledge of the life cycle of their type of raspberry plants to develop a management plan. He asks students to reflect on what they have noticed so far about how the plants have changed since they started working in the greenhouse several months ago. Students work in groups to discuss what they have recorded in their log books and develop an idea about the stages of the life cycle. Some students have experience growing raspberries with their families at home, so they contribute those ideas to the discussion. Each group presents their ideas, and Dr. Webb helps the class combine their ideas into a whole-class model. He then connects what they have constructed to the scientifically accepted model, helping them deepen their understanding and give names to what they have observed. They then use this knowledge to help decide what their next steps will be in the greenhouses after the harvest is complete.

This vignette illustrates how Dr. Webb uses experiential learning to support student understanding of processes and procedures through providing ongoing concrete real-world experiences, guiding students in reflecting on these experiences, helping them make meaning of their experiences through connecting to theoretical concepts, and providing opportunities to apply their new understandings in authentic practice. This ongoing cycle helps him prepare students to be able to flexibly apply their knowledge in the future. Considerations for utilizing production experiences in the laboratory environment as the basis for experiential learning are as follows:

Concrete Experience

Agricultural education laboratory environments often mirror the conditions of real-world facilities that students will encounter in industry. Teachers of agriculture use the laboratory environment to engage their students in real-world tasks related to production, such as harvesting, processing, and cleaning and maintenance of facilities. These tasks are often part of long-term, client-facing efforts, such as a plant sale or timber harvest, that raise funds for the program. These tasks, while aligned with the overall course, are often ongoing and may or may not be directly tied to planned instruction on a particular day.

Through these concrete experiences, students make new observations, apply critical thinking and problem-solving skills in practice, and build up a range of experiences that become a rich foundation for deeper learning. Designing meaningful agricultural production experiences requires ensuring that the laboratory environment is accessible and welcoming to all students. Teachers do this through using their knowledge of the skills and abilities their students have and of the students' prior experience in agriculture as well as through laboratory management strategies to assign tasks and support students.

Closing the Loop of the Experiential Learning Cycle

As with all experiential learning, the stages of reflective observation, abstract conceptualization, and active experimentation are important considerations in instructional design. In reflective observation, students take notice of what they experienced; they are guided to take a step back from the experience itself and consider what might be important about what occurred. In abstract conceptualization, students start to draw more general conclusions from what they have noticed through reflective observation. Through support from the teacher, they may also make connections between what they experienced and concepts they have learned in the classroom to help understand what they observed. In active experimentation, students apply their new understandings in practice. These stages can happen through conversations during practice in the laboratory environment or more formally in the classroom. In many cases, the active experimentation stage serves as a new experience to start the cycle over again.

Check for Understanding

Review the vignette describing Dr. Webb's agricultural production class. Can you identify how he supported experiential learning?

Designing Discovery Learning for Coconstruction of Solutions to Real-World Problems

Discovery Learning in Action at CCHS

In their agricultural biotechnology class, Dr. Randy Webb and Ms. Rachelle Rasco use the design-based learning model (Wells, 2021) to engage students in a semester-long capstone project. Teams of students choose a real-world problem that relates to their career interests. They use techniques they learn in the biotechnology lab and knowledge they gain from field trips and in the classroom in order to design and test a solution. Dr. Webb and Ms. Rasco don't know what the answers are, but they are there to support the teams along the way, providing guidance and resources as they need them. This semester, one of the teams is trying to figure out how you could feed a cow in space for one year! They are experimenting with techniques for growing crops in agar to alleviate the need to water them and address the gravity issue by holding the plants in place. They are testing out what concentrations of seeds, agar, and fertilizers would work best. Their initial solutions failed, but they are rethinking their process and preparing to redesign their solution and retest it. Their teachers are there to help them think about what went wrong and ask good questions to help them move forward. They also share their progress with the other groups in the class so that they can learn from each other. Confident that they will have something interesting to share no matter what happens next, the group is planning to write a research report, create a research poster, and present their findings to local industry leaders, news media, and school administrators. (Note: This description is excerpted from Webb et al., 2020.)

This vignette illustrates how Dr. Webb and Ms. Rasco use design-based learning (Wells, 2021) to engage students in developing their own solutions to real-world problems. In this process, students deepen their understanding of course content and practice higher-order thinking skills. The learning and application of content knowledge is driven by the problem students are trying to solve and the teachers serve as a guide in the process. Giving students voice and choice in their learning though this mode supports motivation and connection to identity. Considerations for designing discovery learning in the laboratory environment are as follows:

Choosing an Instructional Model

Unlike the other two modes discussed in this chapter, there are many potential instructional models that are grounded in constructivism and promote discovery learning. Design-based learning is one. Other common models are problem-based learning and the scientific-inquiry learning cycle.

Problem-based learning originated in the field of medical education as a way to train medical professionals for the real-world decision-making and problem solving that they would encounter on the job (Barrows & Tamblyn, 1980), and it has been adopted widely. In this model, students are presented with a real-world problem and the learning occurs through their attempts to determine a solution and defend their reasoning. A common form of the scientific-inquiry learning cycle (as described in NRC, 2000) is the BSCS 5E Instructional Model. This model has five stages (engagement, exploration, explanation, elaboration, and evaluation) that direct students to answer a scientifically oriented question through hands-on exploration using authentic scientific practices (NRC, 2012).

Resources in the Laboratory Environment

Leveraging the resources of the laboratory environment in design of discovery learning, either entirely or as part of the experience, requires careful consideration of how and why students will engage with them. Typically, this means making a variety of tools and materials available to students to use as their needs emerge in process. Building in a proposal process in which teachers can review plans and approve the use of materials and space can help in management. For example, in a scientific-inquiry unit where students are designing experiments with plants in the greenhouse, groups can submit their proposed methods and list of materials in advance. This allows the teacher to gather what students need and review any safety concerns prior to setting up the experiment. In other cases, it may be appropriate to limit the materials and equipment available to the students. For example, in a woodworking design challenge, the teacher can give each group a standard set of materials to work with and limit them to using hand tools. Groups would have creativity in how they meet the design brief, but would also be guided to practice a specific set of skills.

Cooperative Learning

In the laboratory environment, management of group work is particularly important to make sure that students are engaged and safely using equipment and resources. Cooperative learning strategies are effective in promoting engagement and mediating potential conflicts due to differing social identities (Johnson & Johnson, 1999); this aspect of planning is especially important in longer term projects in which team dynamics can play a big role. Where appropriate, assigning specific roles within a group can help things run smoothly and can reinforce real-world career options within an industry. For example, one student can be in charge of procurement of materials, another can serve as project manager, and another can be the safety officer.

Authentic Assessment

Discovery learning in agricultural education contexts should rely on authentic assessment embedded in the design of the learning experience from the beginning. The laboratory environment provides endless possibilities for students to situate their learning in real-world problem solving. Formative assessment should happen throughout in order to keep groups moving in a productive direction. Summative assessments should include a product that is consistent with the task assigned and, wherever possible, should be shared with other stakeholders outside of the classroom community.

Check for Understanding

Review the vignette describing Dr. Webb and Ms. Rasco's agricultural biotechnology class. Can you identify how they used the laboratory environment, embedded authentic assessment, and connected to student interest?

Learning Confirmation

To check your understanding of the ideas presented in this chapter, please complete the following:

1. Compare and contrast the three modes for utilizing the laboratory environment. In your response, consider the types of student learning goals, theoretical underpinnings, and the role of the laboratory environment.
2. Choose a unit that you are excited to teach that includes the laboratory environment and complete the following:
    a. Using the guidance in chapter 7, identify the learning objectives for your unit.
    b. Determine which mode (or combination of modes) you think best fits your objectives and explain why.
    c. Develop a sketch outline of your unit plan, making sure it includes the objectives, assessment evidence, and a brief description of how you will structure the learning activities.
    d. Consider the mode(s) you didn't choose for this unit. Either
        i. Describe how the unit plan (objectives, assessment evidence, and learning activities) would be different if you utilized these different modes;
        OR

ii. Create new sketch unit outlines (objectives, assessment evidence, and learning activities) for your unit using each of the different modes.

3. Within each of your unit plan ideas from number 2, complete the following:
    a. Identify potential challenges that might arise when teaching the unit and how you will go about addressing them.
    b. Describe ways in which you will attend to diversity and inclusion to best meet the needs of all of your students when teaching the unit.
    c. Reflect on what excites you most about the potential for powerful teaching and learning within the unit.

For additional practice, consider these additional assessment tasks:

1. Swap unit plan outlines with a classmate and give them advice for how they could strengthen use of the laboratory environment by considering a different mode.
2. Interview an experienced agriculture teacher about how and why they utilize the laboratory in their program. After the interview, identify which mode you think best fits their practice. Defend your answer with evidence from the interview.

Applying the Content

Your Role as a Designer

In many fundamental ways, teachers are designers. Working within existing curriculum frameworks and the real-world constraints of their educational environments, teachers make design decisions every day as they plan for how they will support student learning in powerful ways. Just like other design tasks, there is not a one-size-fits-all solution—how a teacher applies the ideas presented in this chapter will depend on their context and the laboratory environment(s) available for use. The different modes can be considered as potential solutions to instructional design challenges; they are offered as a starting point for decision-making about how and when to use the laboratory environment to support student learning. As teachers consider the resources available in the laboratory environment and/or new facilities and resources they would like to acquire, student learning is front and center. Identifying the learning goals for a course or unit and how these are situated in real-world contexts is a great starting point for determining which mode of instruction will be most suitable.

Reflective Questions

1. Laboratory management and safety concerns require the teacher to do a lot of detailed planning and careful coordination of the daily activities of students. Why do you think the author of this chapter decided to focus instead on modes for teaching in laboratory environments?
2. Recall the laboratory environment that you pictured at the beginning of this chapter. In what ways are you thinking about using that space to support student learning that you hadn't considered before you read this chapter? What ideas did you have that were reinforced through the ideas presented in this chapter?
3. Throughout this chapter, fundamental ideas that are introduced in other parts of this textbook were reinforced or referenced. Revisit your responses to the assessment questions in the chapters in part 3 (7, on planning; 8, on delivery; and 9, on evaluation). In what ways did your responses include the laboratory environment? In what ways should you expand your responses to make sure that you carefully consider the laboratory environment as you design instruction?

Glossary of Terms

- **agriculture laboratory environments:** Instructional spaces in which students are using authentic resources in hands-on ways to support learning about agriculture
- **backward design:** Designing learning with the end in mind; an approach to instructional planning that starts with identifying learning goals, proceeds to determining what evidence of learning is needed, and concludes with planning learning experiences
- **direct instruction:** Teacher-centered instructional mode that scaffolds skill development as students work towards independence with teacher guidance
- **discovery learning:** Umbrella term for student-centered instructional models grounded in inductive reasoning, such as problem-based learning, design-based learning, and inquiry-based learning
- **experiential learning:** Instructional mode based on Kolb's experiential learning theory that centers learning from reflection on concrete experience
- **inductive reasoning:** Going from specific to general; building abstract understanding from concrete experience
- **psychomotor skill:** task that depends on coordination of both cognitive (mind) and motor (body) functions

References

Barrows, H. S., and R. M. Tamblyn. (1980). *Problem-based learning: An approach to medical education*. Springer.

Dell'Olio, J. M., & Donk, T. (2007). *Models of teaching: Connecting student learning with standards*. SAGE Publications, Inc. https://doi.org/10.4135/9781452232324

Hunter, M. (1982). *Mastery teaching*. TIP Publications.

Johnson, D., & Johnson, R. T. (1999). Making cooperative learning work. *Theory Into Practice, 38*(2), 67–73. https://doi.org/10.1080/00405849909543834

Kolb, D. A. (2015). *Experiential learning: Experience as the source of learning and development* (2nd ed.). Pearson Education LTD.

National Research Council (NRC). (2000). *Inquiry and the national science education standards: A guide for teaching and learning.* National Academy Press.

National Research Council (NRC). (2012). *A framework for K-12 science education: Practices, crosscutting concepts, and core ideas*. The National Academies Press.

Osborne, E. (1986). Teaching strategies for developing psychomotor skills. *NACTA Journal, 30*(1), 54–57. https://www.jstor.org/stable/43755313

Schunk, D. H. (2012). *Learning theories: An educational perspective* (6th ed.). Pearson.

Webb, R., Rasco, R., Steger, D., & Scherer, H. H. (2020). Students Discover Career Ready Skills Through Biotechnology. *The Agricultural Education Magazine, 93*(1), 45–48. https://www.naae.org/profdevelopment/magazine/archive_issues/Volume93/2020%2007%20-%20July%20August.pdf

Wells, J. G. (2021). *Design-based biotechnological learning: Distinct knowledge forms supporting technology and science conceptual understanding*. In Design-based concept learning in science and technology education, edited by Ineke Henze and Marc J. de Vries, 223–47. Koninklijke Brill NV. https://doi.org/10.1163/9789004450004_011

APPENDIX 12A: IN-SCHOOL AGRICULTURE LABORATORY ENVIRONMENTS

AFNR Career Pathway	Example Laboratory Environments
Agribusiness Systems	Customer service area Computer lab
Animal Systems	School barns Animal clinic Grooming area Kennels
Environmental Service Systems	Water testing lab Soil testing lab Computer lab School grounds and facilities
Food Products & Processing Systems	Food preservation lab Meat processing facility Microbiology lab
Natural Resources Systems	Nature trail Outdoor lab School grounds Wetland Forest
Plant Systems	School farm School grounds Greenhouse Turf plots Hydroponics lab
Power, Structural & Technical Systems	Ag mechanics learning laboratory School farm shop Computer lab

In-school agriculture laboratory environments.

APPENDIX 12B: INTRODUCTORY CONSIDERATIONS FOR LABORATORY SAFETY AND MANAGEMENT

Providing for Safety Instruction

An essential element of good laboratory management is providing for safety instruction. Every agricultural laboratory has many dangerous areas and situations. It is crucial that students learn to work in their environment safely not only for their immediate welfare, but so that they develop the essential safety habits needed for their future employment in the industry. From the standpoint of teachers, there are at least three compelling reasons for making sure that their students learn to work safely in the laboratory. First of all, teachers care about the individual students and certainly would not want to see one injured. Second, teachers want to get students totally prepared for a successful future. The third reason teachers must be sure students receive superb safety instruction is to protect their own welfare. The legal climate is such that teachers who do not provide satisfactory safety instruction and supervision may be found liable for failure to do so. Thus, teachers must be sure they are adequately covered by reliable professional liability insurance.

The first thing to remember about safety instruction is that safety is largely a question of attitude. Thus, teachers must use instructional practices that impact not only the psychomotor and cognitive domains but also impact favorably on the affective domain.

The cornerstone for safety instruction is thorough classroom instruction. Students must be taught the specific safety practices that will be expected and required in the laboratory in question. Students must also be taught specific safety practices that pertain to each area of the lab and to specific pieces of equipment and tools.

Structured classroom instruction with complete note taking is essential. Teachers should use a variety of techniques that appeal to all five senses when teaching safety principles and practices. Another effective technique is to bring in people from agribusiness and industry who can relate vivid experiences they have had or seen and to offer industry specifications on how safety is handled in real-life settings.

Teachers also need to demonstrate and role-model the specifics of safe psychomotor operations. Finally, students must show "proficiency" on a general safety test, as well as a specific test for each category of tools and equipment, chemicals and dangerous agents, and specific learning centers. These tests then must be kept on file as evidence that formal instruction has taken place and that each student has demonstrated mastery. There are several good commercial sources of safety units, tests, and audiovisual aids.

Then, throughout the duration of the course, students must be reminded of safety. As new problems are discussed, teachers can reiterate previous cautions.

Once in the laboratory, there needs to be an effective, safety-conscious environment. Teachers must use state and national recommendations for color-coding, display safety posters and exhibits (many good materials are available through each state's safety agencies), and guarantee that all guards and other safety devices on equipment meet state standards. The laboratory must always be neatly arranged, clean, and well lit.

Each day and throughout each laboratory period students need to be reminded of safety. This can be done by the teacher, by the laboratory supervisor, and by the safety engineer if one is used. Another important practice is for teachers to set the example by wearing safe clothing and safety glasses and following the same rules that they have set forth for their students.

The final element of a good safety program is for the teacher and assigned students to conduct safety inspections daily. If tools and equipment are found to be in anything other than top operating condition, appropriate restorative maintenance must be provided. All in all, the goal of a good program of safety instruction is for students to develop a mind set for safety and practice daily habits of working safely.

Managing Tools and Equipment

Teachers must be sure to manage all tools and equipment, including computers. Without a viable management system for maintaining and upgrading tools and equipment, there will not be a laboratory to manage. Tools, equipment, hardware, and software are very expensive, and it is the teacher who is ultimately responsible for keeping them in good condition or updated. Also, if tools, equipment, and computers are in disrepair, broken, or missing, the students will be unable to stay productively involved during laboratory, the consequences of which can be readily predicted.

If students do not learn to properly use and care for tools, equipment, hardware, or software as a part of their instruction, then they will encounter serious problems when they enter the work force. In essence, students need to develop acceptable equipment use habits. The dollars involved in industry demand such an attitude.

Many teachers prefer to store or locate tools of all kinds on a display board near the location where they will be used. Larger tools and pieces of portable equipment should be kept in a logical storage area and checked in and out using a system that everyone understands. Wherever tools are kept, they should be kept clean and ready for the function for which they were intended.

This means that every student who uses a tool or piece of equipment returns it cleaned and ready to use to the place where they found it. At the close of every laboratory period, the supervisor and teacher must be certain that everything is clean and in its place and, in the case of shared computers, certain that no personal files are left on hard drives. Otherwise, problems mount.

Teachers must also provide students with instruction on refitting tools, upgrading software packages, and fine-tuning equipment. For example, students need to learn to sharpen hoes, shovels, chisels, chain saw chains, and other tools. They must learn to replace glass or plastic on the green house, readjust gate hinges in the school barns, sharpen clippers used for dog grooming, and re-adhere PVC pipe on aquaculture tanks. This is an important part of preparation for work and is vital to the successful management of tools and equipment.

Procedures for Efficiently Starting and Ending a Laboratory Period

Beginning the laboratory period. If teachers begin laboratories in a businesslike manner, then this mood is apt to prevail for the rest of the laboratory period. However, if teachers begin laboratories in a disorganized and confused manner, then the entire laboratory period is apt to be tainted by that approach and attitude throughout the session.

The first thing to be accomplished in getting the laboratory started is to achieve the proper mental set. The mental set that is desired is that "we're here to work, we'll accomplish much, and we'll enjoy doing it." This can be achieved by beginning the laboratory promptly and in a systematic and businesslike fashion, preferably in the classroom where students can sit in the academic environment and check notebooks for project timeliness and daily objectives before reporting to the laboratory space.

Students need to report for lab promptly and be dressed and ready for work within a set number of minutes after being dismissed from the classroom. The roll should then be checked in the laboratory environment to ensure each student is where they need to be for that day. This can be done by the teacher or by the laboratory supervisor for that day.

Assignments for the day then need to be given. They may be given on paper, on the chalkboard, or orally. Each student must know their assignment. The teacher needs to be sure that the students know the goals or objectives for the lab activities and that they clearly see the relationship of their lab work to their previous classroom learning.

Prior to allowing students to begin work, the teacher or an appointed student, such as the laboratory supervisor for that day, needs to double-check to be sure that every student knows what to do and how to get started. Only then should students be allowed to start work.

If this procedure or a similar one is not followed, then some students begin work, others do not, there is more opportunity for off task behavior, there is confusion as to who is supposed to do what, and the general atmosphere of the laboratory is not conductive to learning.

Procedures to follow in ending the laboratory period. Just as the laboratory period needs to be started in a businesslike manner, it is also imperative that it be ended in such a manner. The following procedures are suggested:

1. *Signal a time to stop.* This may be done by blowing a whistle, ringing a bell, or other procedures that can become habitual for students. Upon giving the "stop work" signal, everyone must immediately cease work; otherwise, the system does not work. This expectation has to be clearly defined early in the school year, reinforced, and practiced.
2. *Student put away their work, tools, and equipment.* Students need to clear the laboratory so it can be cleaned daily. Projects cannot be left strewn about. Students cannot be allowed to leave tools out, reasoning that they'll be needed again tomorrow or by students in another laboratory. Such practice leads to a situation in which no one can find anything. If a student is acting as the equipment manager, all items used in the laboratory were checked-out by that person and must be checked back in by that person.
3. *Everyone joins in to clean the entire laboratory.* The only way this works is if everyone has a specific duty during clean-up. Teachers ought to demonstrate how they want each clean-up job performed; otherwise, perfection will not be attained. These duties need to be rotated since some are much less pleasant than others. Many teachers in agriculture mechanics laboratories use a clean-up wheel, which can be created for all types of laboratories. All students' names in the class are included in the center circle. Then one clean-up duty for each name is listed on the outer circle. The wheel rotates clockwise one name each class period.
4. *Once all clean-up duties have been completed, students may clean themselves up and dress for the rest of their school day.* This prioritizes the clean-up of the collective space first and ensures that it is complete.
5. *Students wait for dismissal from the teacher.* Once students are ready to leave the laboratory they must wait until the teacher is satisfied, primarily via the laboratory supervisor, that everything is in order. The teacher then needs to dismiss the class. The students do not leave until the teacher has dismissed them; otherwise, laboratory clean-up may not be completed well.

If the laboratory is not cleaned after every class period, then no system will work. A clean laboratory is an efficient and productive laboratory and must be a priority item with teachers and students alike.

Using Learning Centers to Manage Laboratory Tasks with Limited Materials

Generally, there is not sufficient work, materials, or equipment that all the students in the laboratory can be doing the same thing. For example, if students need to practice transplanting shrubs, seldom are there enough shovels or pruners and other equipment to accommodate everyone in the class at the same time. In the case of an animal care program, a school generally does not have enough aquaria so that each student or even pair of students can simultaneously master the process of setting up an aquarium. Thus, teachers have to design what are called learning centers in order to have all of the students productively involved.

Setting up learning centers. A learning center is an area (physically and in terms of subject matter) within the laboratory where similar (like) work is performed. The specific learning centers in a laboratory during a given part of the school year are determined by the teacher's course of study. While the ideal of presenting the basics in the classroom, followed immediately by practice, is never perfectly attained, there must be a reasonable degree of connection between the development of understandings and the development of skills.

Once learning centers have been identified or chosen, the teacher must plan the specific learning activities to be completed in each of the learning centers. This should not be construed to mean that every group of students who work in a learning center completes the same activities. For example, the group that works in the flower shop one week might make three planters, but the group the next week may not make any. It all depends on the work that needs to be done during the time students are in the learning center. However, if the teacher is rotating students through a learning center to learn striking an arc and laying a bead in welding (initial skill development), then all students who go through that center will complete the same activities since this is a basic skill development-learning center. This planning is fairly well dictated by the work to be done and the prior instruction that students have had. Before students can engage in meaningful practice, they must have a precise idea of what they are to do, what tools and supplies they need in order to do it, and what procedures they must follow. Once teachers develop assignment sheets for the basic learning center assignments, this part of planning can be used and reused.

Rotating students through learning centers. If all the students cannot do the same tasks at the same time and must divide into work groups and work at separate learning centers during lab, then the teacher must have a viable system to make sure that over a period of time, all students get all of the essential experiences. This plan is called a rotation system. It is a plan that ensures that each work group of students has a chance to work through all the learning centers over a period of time. Depending on the nature of the program, this rotation could be completed once during the school year or could be completed every week or two.

For example, in horticulture a group of students might work in the greenhouse for one day, work in the flower shop the next, care for a school golf green the third day, prune trees the fourth day, and maintain horticulture equipment the final day of the week. However, in an agricultural industrial mechanics program a group of students may work on engine overhaul for an entire grading period. The length of time in a center, the number of centers, and the frequency of rotation from one center to another depends on the nature of the work being done and the scope of the work. It takes longer to assemble an aquaculture system than it does to make a corsage and therein lies the determiner of frequency of rotation.

Rotation is not the easiest way to manage laboratory learning. Typically, the easiest way to manage a laboratory is to have every student practicing the same skill. Nor is rotation the most preferred way to manage a laboratory based on educational soundness. Rather it is a necessity. However, rotation is far better than planning enough lab work for each individual student for one day, and then the next day starting to plan from scratch all over again. Its primary strength is that it brings an acceptable degree of systematization, clarity, and structure. It also provides for accountability in that it allows every student to work in every area.

A rotation schedule is planned by grouping work into a number of areas that can accommodate the number of students in the class in mathematically divisible units. For example, if one has twenty-four students, then four learning centers that accommodate six students each might be used; or six centers of four each; or two centers for six students each and four centers for three students each. Another important consideration is that the learning centers require about the same amount of time or divisible portions thereof. For example, if there are four centers of six students each, then all four centers need to require the same amount of time (one week, one month, or whatever is needed). In the case of using two groups of six each and four groups of three each, the groups of three would have to have twice the number of stations as the groups of six, that is, groups of six spend one month (or other unit of time) in a center while the groups of three work in two stations of two weeks each.

Note

The material in this appendix has been reproduced under fair use with slight modification from the original.

Newcomb, L. H., McCracken, J. D., Warmbrod, J. R., & Whittington, M. S. (2004). *Methods of teaching agriculture* (3rd ed.). Prentice Hall.

Made in the USA
Monee, IL
21 July 2025

21572247R00151